Green Energy and Technology

Ayhan Demirbas

Biorefineries

For Biomass Upgrading Facilities

Springer

Ayhan Demirbas, Professor of Energy Technology
Sila Science and Energy
Trabzon
Turkey

ISSN 1865-3529
ISBN 978-1-4471-2516-7 e-ISBN 978-1-84882-721-9
DOI 10.1007/978-1-84882-721-9
Springer Dordrecht Heidelberg London New York

British Library Cataloguing in Publication Data
A catalogue record for this book is available from the British Library

Cover design: WMXDesign, Heidelberg, Germany

Printed on acid-free paper

Springer is part of Springer Science+Business Media (www.springer.com)

Preface

A biorefinery is a facility that integrates biomass conversion processes and equipment to produce fuels, power, and value-added chemicals from biomass. The biorefinery concept is analogous to today's crude oil refineries, which produce multiple fuels and products from petroleum. The goal in biorefining is to isolate all the added value from the biomass feedstock, resulting in little or no waste. This will not only improve the economics so that such processes can compete with the petrochemical industry, but will also lower the overall environmental impact. A biorefinery can also be defined as a facility that produces food, feed, materials, chemicals, power and/or heat from biomass. By producing multiple products, a biorefinery can take advantage of the differences in biomass components and intermediates, and maximize the value derived from the biomass feedstock. Biofuels appear to be a "greener" energy substitute for fossil fuels. Since biofuels are renewable, carbon neutral, and available throughout the world, they could contribute to sustainable development and global environmental preservation.

The production of liquid hydrocarbon fuels for the transportation sector is of importance for continued vitality of our industrialized society. Research is being conducted worldwide to develop new technologies for the generation of liquid fuels from renewable resources. Currently, biodiesel is produced by transesterification of vegetable oils and animal fats, and ethanol by fermentation of glucose. Processes for the efficient gasification of biomass to produce CO and H_2 (synthesis gas or syngas) are being developed. The syngas can be further processed to produce methanol or liquid alkanes by Fischer–Tropsch synthesis using well-established industrial processes.

The biorefinery economy is a vision for a future in which biorenewables replace fossil fuels. The transition to a biorefinery economy would require a huge investment in new infrastructure to produce, store and deliver biorefinery products to end users.

Policy drivers for renewable liquid biofuels have attracted particularly high levels of assistance in some countries given their promise of benefits in several areas of interest to governments, including agricultural production, greenhouse gas emissions, energy security, trade balances, rural development and economic opportunities for developing countries.

Current energy policies addressing environmental issues include environmentally friendly technologies that increase energy supplies and encourage cleaner, more efficient energy use, and also air pollution, the greenhouse effect, global warming, and climate change.

There are some barriers to the development of biofuel production, including technological, economical, supply, storage, safety, and policy barriers. Reducing these barriers is one of the driving factors in the government's involvement in biofuel and biofuel research and development. Production costs are uncertain and vary with the feedstock available. The production of biofuels from lignocellulosic feedstocks can be achieved through two very different processing routes: biochemical and thermochemical. There is no clear candidate for "best technology pathway" between the competing biochemical and thermochemical routes. Technical barriers for enzymatic hydrolysis include: low specific activity of current commercial enzymes, high cost of enzyme production, and lack of understanding of enzyme biochemistry and mechanistic fundamentals.

There are many technical and non-technical barriers related to the implementation and commercialization of the biorefinery. Current technical barriers with the use of energy crops are associated with the cost of production and difficulties in harvesting and storing the material grown, especially for crops that have to be harvested within a narrow time period in the autumn. Transportation costs are of prime importance when calculating the overall costs of biomass; hence, local or regional production of biomass is most favorable.

This book on biofuels attempts to address the needs of energy researchers, chemical engineers, chemical engineering students, energy resources specialists, agriculturists, crop cultivators and others interested in a practical tool for pursuing their interests in relation to bioenergy. Each chapter in the book starts with fundamental explanations suitable for a general audience and ends with in-depth scientific details for expert readers. General readers will include people interested in learning about solutions for current fuel and environmental crises. The expert readers will include chemists, chemical engineers, fuel engineers, agricultural engineers, farming specialists, biologists, fuel processors, policy makers, environmentalists, environmental engineers, automobile engineers, college students, research faculty, etc. The book may even be adopted as a textbook for college courses that deal with renewable energy and/or sustainability.

Trabzon, Turkey (April, 2009) Ayhan Demirbas

Contents

Chapter 1
Introduction

1.1 Introduction

Human development has always been associated directly or indirectly with energy use, and because of this the sources of energy were dealt with in the recent past as fully available to human necessity. The importance of energy as an essential ingredient in economic growth, as well as in any strategy for improving the quality of human beings, is well established. Energy is fundamental for the social and economic development and stability of any country. Reality shows that the energy availability from non-renewable sources is limited, and the exploration, processing and the use of that energy seriously impact the environment.

Energy sources are classified into three groups: fossil, renewable and fissile. The term fossil refers to an earlier geologic age, and fossil fuels, formed a great many years ago, are not renewable. The fossil energy sources are petroleum (crude oil), coal, bitumen, natural gas, oil shales, and tar sands. The main fissile energy sources are uranium and thorium. Figure 1.1 shows the fossil energy reserves of the world. Coal reserves of the world are 3.2×10^{17} MJ. On the other hand, deuterium and uranium reserves of the world are 7.5×10^{24} and 1.2×10^{20} MJ, respectively. About 98% of carbon emissions result from fossil fuel combustion. Reducing the use of fossil fuels would considerably reduce the amount of carbon dioxide and other pollutants produced. This can be achieved by either using less energy altogether or by replacing fossil fuels with renewable fuels.

Petroleum is the largest single source of energy consumed by the world's population, exceeding coal, natural gas, nuclear sources and renewables, as shown in Fig. 1.2 for the year 2005. In fact today, over 80% of the energy we use comes from three fossil fuels: petroleum, coal, and natural gas. Unfortunately, petroleum oil is in danger of becoming short in supply. Hence, the future trend is towards using alternate energy sources. Fortunately, technological developments are making the transition possible.

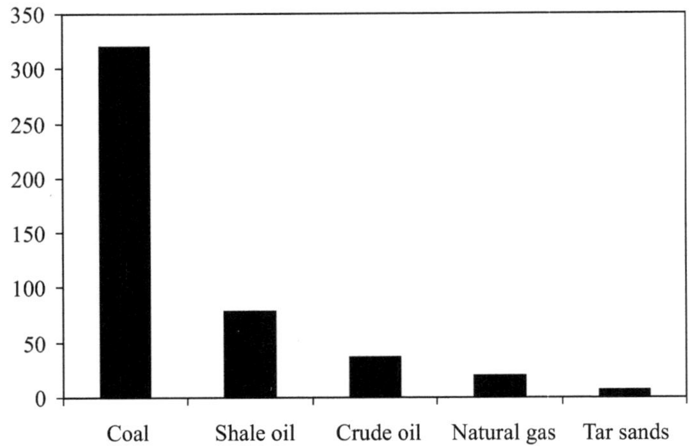

Fig. 1.1 Fossil energy reserves of the world. Each unit = 1×10^{15} MJ = 1.67×10^{11} Bbl crude oil

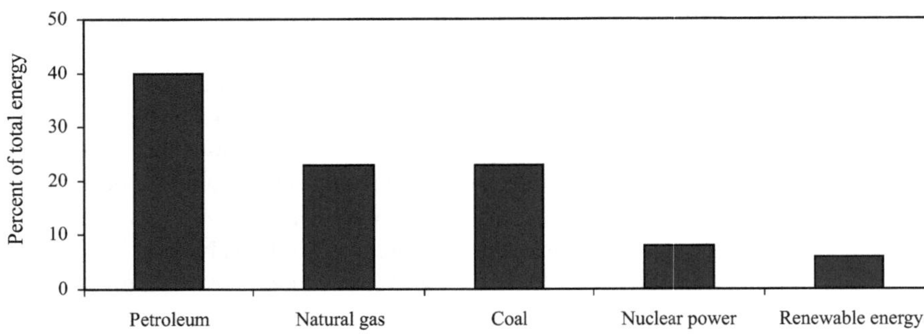

Fig. 1.2 Global energy consumption for 2005

About 98% of carbon emissions result from fossil fuel combustion. Reducing the use of fossil fuels would considerably reduce the amount of carbon dioxide and other pollutants produced. This can be achieved by either using less energy altogether or by replacing fossil fuel by renewable fuels. Hence, current efforts focus on advancing technologies that emit less carbon (e. g., high-efficiency combustion) or no carbon such as nuclear, hydrogen, solar, wind, geothermal, or on using energy more efficiently and on developing sequestering carbon dioxide emitted during fossil fuel combustion (Demirbas 2009).

Another problem with petroleum fuels is their uneven distribution in the world; for example, the Middle East has 63% of the global reserves and is the dominant supplier of petroleum. This energy system is unsustainable because of equity issues as well as environmental, economic, and geopolitical concerns that have far reaching implications. Interestingly, the renewable energy resources are more evenly distributed than fossil or nuclear resources. Today's energy system is unsustainable because of equity issues as well as environmental, economic, and geopolitical concerns that will have implications far into the future. Hence, sustainable renewable energy sources such as biomass, hydro, wind, solar (both thermal and photovoltaic), geothermal, and marine energy sources will play an important role in the world's future energy supply. For example, it is estimated that by year 2040 approximately half of the global energy supply will come from renewables, and the electricity generation from renewables will be more than 80% of the total global electricity production (Demirbas 2009).

1.2 Short Supply of Petroleum

The word *petroleum* comes from the Greek word *petra*, meaning rock, and the Latin word *oleum*, meaning *oil*. Oil is a thick, dark brown or greenish liquid found in reservoirs in sedimentary rock. Tiny pores in the rock allow the petroleum to seep in. These "reservoir rocks" hold the oil like a sponge, confined by other, non-porous layers that form a "trap." It is used to describe a broad range of hydrocarbons that are found as gases, liquids, or solids beneath the surface of the Earth. The two most common forms are natural gas and crude oil. Petroleum consists of a complex mixture of various hydrocarbons, largely of the alkane and aromatic compounds. The color ranges from pale yellow through red and brown to black or greenish. Petroleum is a fossil fuel because it was formed from the remains of tiny sea plants and animals that died millions of years ago, and sank to the bottom of the oceans.

According to the most widely accepted biogenic theory, crude oil, like coal and natural gas, is the product of compression and heating of ancient vegetation and animal remains over geological time scales. According to this theory, organic matter is formed from the decayed remains of prehistoric marine animals and terrestrial plants. Over many centuries this organic matter, mixed with mud, was buried under thick sedimentary layers. The resulting high pressure and heat causes the remains to transform, first into a waxy material known as kerogen, and then into liquid and gaseous hydrocarbons by the catagenesis process. The fluids then migrate through adjacent rock layers until they become trapped underground in porous rocks known as reservoirs, forming an oil field, from which drilling and pumping remove the liquid. The reservoirs are at different depths in different parts of the world, but the typical depth is 4–5 km. The thickness of the oil layer is about 150 m and is generally termed the "oil window." Three important elements of an oil reservoir are: a rich source rock, a migration conduit, and a trap that forms the reservoir (Demirbas 2009).

The earliest known crude oil wells were drilled in China as early as the fourth century. The wells, as deep as 243 m, were drilled using bits attached to bamboo poles. The oil was burned to produce heat needed in the production of salt from brine evaporation. By the tenth century, extensive bamboo pipelines connected oil wells with salt springs. Ancient Persian tablets indicate the medicinal and lighting uses of petroleum in the upper echelons of their society. In the ninth century, oil fields were exploited to produce naphtha in Baku, Azerbaijan. These fields, first described by the geographer Masudi in the tenth century, had increased output to hundreds of shiploads in the thirteenth century as described by Marco Polo.

The modern history of *petroleum* began in 1846, with the discovery of the refining kerosene from coal by Atlantic Canada's Abraham Pineo Gesner. Poland's Ignacy Łukasiewicz discovered a means of refining kerosene from the more readily available "rock oil" (petroleum) in 1852, and in the following year the first rock oil mine was built in Bobrka, near Krosno in southern Poland. The discovery rapidly spread around the world, and Meerzoeff built the first Russian refinery in the mature oil fields of Baku in 1861, which at the time produced about 90% of the world's oil. In fact, the battle of Stalingrad was fought over Baku (now the capital of the Azerbaijan Republic).

The first commercial crude oil well in North America was drilled by James Miller Williams in 1858 in Oil Springs, Ontario, Canada. In the following year, Edwin Drake discovered oil near Titusville, Pennsylvania. There, he pioneered a new method for producing oil from the ground, in which he drilled using piping to prevent borehole collapse, allowing for the drill to penetrate deeper into the ground.

For about 10 years Pennsylvania was the one great crude oil producer of the world, however since 1870 the industry has spread all over the globe. From the time of the completion of the first flowing well on the Baku field to now, Russia has ranked second on the list of producing countries, while the region of Galicia in Central Europe and Romania became prominent in 1878 and 1880, respectively. Sumatra, Java, Burma, and Borneo, where active development began in 1883, 1886, 1890, and 1896, respectively, are fast becoming the chief sources of the oil supplies of the world.

Before the 1850s, Americans often used whale oil to light their homes and businesses. Drake refined the crude oil from his well into kerosene for lighting, which was used till the discovery of light bulbs. Gasoline and other products made during refining were simply discarded due to lack of use. In 1892, the "horseless carriage" solved this problem since it required gasoline. By 1920 there were nine million motor vehicles in the US and many gas stations to supply gasoline.

Petroleum is the most important energy source, as 35% of the world's primary energy need is met by crude oil, 25% by coal, and 21% by natural gas, as shown in Table 1.1 (IEA 2007). The transport sector (i.e., automobiles, ships, and aircraft) relies well over 90% on crude oil. In fact, the economy and lifestyle of industrialized nations rely heavily upon sufficient supply of crude oil at low cost.

There are two great petroleum organizations in the world: OPEC and OECD. OPEC is the Organization of Petroleum Exporting Countries and its current members include Iran, Iraq, Kuwait, Saudi Arabia, Venezuela, Qatar, Indonesia, Libya, United Arab Emirates, Algeria, Nigeria, Ecuador and Gabon. The OPEC members try to set production levels for petroleum to maximize their revenue. According to supply/demand economics, the less oil they produce, the higher the price of oil on the world market and the more oil they produce, the lower the price. However, the OPEC countries do not always agree with each other. Some OPEC countries want to produce less oil to raise prices, whereas the other OPEC countries want to flood

Table 1.1 Year 1973 and 2005 fuel shares of total primary energy supply (TPES) (excludes electricity and heat trade)

	World		OECD	
	1973	2005	1973	2005
Oil	46.2	35.0	53.0	40.6
Coal	24.4	25.3	22.4	20.4
Natural gas	16.0	20.7	18.8	21.8
Combustible renewables and wastes	10.6	10.0	2.3	3.5
Nuclear	0.9	6.3	1.3	11.0
Hydropower	1.8	2.2	2.1	2.0
Other (geothermal, solar, wind, heat, etc.)	0.1	0.5	0.1	0.7
Total (million tons oil equivalent)	6128	11 435	3762	5546

the market with petroleum to reap immediate returns. In addition, the oil supply can be controlled for political reasons. For example, the 1973 OPEC oil embargo was a political statement against the US for supporting Israel in the Yom Kippur War. Such embargos or cuts in production cause drastic increases in the price of petroleum.

The US is a member of the Organization for Economic Cooperation and Development (OECD), which is an international organization of thirty countries that accept the principles of representative democracy and a free market economy. In the 1970s, as a counterweight to OPEC, OECD founded the International Energy Agency (IEA), which is regarded as the "energy watchdog" of the western world and is supposed to help to avoid future crises. IEA provides demand and supply forecasts in its annual World Energy Outlook (WEO) report, and the current situation of the oil market in its monthly publication. The WEO covers the forecast for the next two decades and is highly regarded by people related to the energy industry.

The price of a barrel (42 gallon or 159 L) of crude oil is highly dependent on both its grade (e. g., specific gravity, sulfur content, viscosity) and location. The price is highly influenced by the demand, current supply, and perceived future supply. Both demand and supply are highly dependent on global macroeconomic and political conditions. It is often claimed that OPEC sets a high oil price, and the true cost of oil production is only $2 per barrel in the Middle East. These cost estimates ignore the cost of finding and developing the oil fields. In fact, the price is set by the cost of producing not the easily extractable oil, but the more difficult marginal oil. For example, by limiting production the OPEC has caused development of more expensive areas of production such as the North Sea.

Oil production in the US peaked at 9.5 million barrels per day in 1970. However, the US petroleum production has declined every year since 1985, and in 2005 it was at 5.2 million barrels per day. The total US consumption of petroleum per day is given in Table 1.2.

Table 1.2 Total US consumption of petroleum per day

Year	Consumption of petroleum (million barrels/day)
1991	5.8
2005	10.1
2007	20.8

In the early 1970s, the inflation-adjusted price hovered near $10/barrel, but it is near $90 and rising irregularly as this book is completed at the end of April 2009. The price of crude oil is reflected in the price of refined products, and gasoline in July 2008 cost as much as $4/gallon in some US markets, and more than $3/gallon on average nationwide.

1.2.1 Petroleum and Heavy Oil Refining

Crude oil (raw petroleum) is separated into fractions by fractional distillation. The fractions at the top are lower than the fractions at the bottom. The heavy bottom fractions are often cracked into lighter, more useful products. All of the fractions are processed further in other refining units.

An oil refinery cleans and separates the crude oil into various fuels and byproducts. The most important one is gasoline. Some other petroleum products are diesel fuel, heating oil, and jet fuel. Refineries use many different methods to make these products. One method is a heating process called distillation. Since oil products have different boiling points, the end products can be distilled or separated. For example, asphalts have a higher boiling point than gasolines, allowing the two to be separated.

Refineries have another job. They remove contaminants from the oil. A refinery removes sulfur from gasoline, for example, to increase its efficiency and to reduce air pollution.

Petrochemicals are chemical products made from raw materials of petroleum or other hydrocarbon origin. Although some of the chemical compounds that originate from petroleum may also be derived from other sources such as coal or natural gas, petroleum is a major source of many.

The two main classes of petrochemical raw materials are olefins (including ethylene and propylene) and aromatics (including benzene and xylene isomers), both of which are produced in very large quantities. At oil refineries, olefins are produced mainly from hydrocarbons by chemical cracking such as steam cracking and by catalytic cracking. At oil refineries, aromatic hydrocarbons are mainly produced by catalytic reforming or similar processes. These basic building blocks make a very wide range of chemicals and other materials used in industry, e. g., monomers, solvents, and adhesives. From the monomers, polymers or oligomers are produced for plastics, resins, fibers, elastomers, certain lubricants, and gels.

World production of ethylene is around 110 million tons per annum, of propylene 65 million tons, and of aromatic raw materials 70 million tons. The largest petrochemical industries are to be found in the US and Western Europe, though the major growth in new production capacity is in the Middle East and Asia. There is a substantial inter-regional trade in petrochemicals of all kinds.

After processing at the refinery, gasoline and other petroleum products are usually shipped out through pipelines. There are about 230,000 miles of pipeline in the US. Pipelines are the safest and cheapest way to move large quantities of petroleum across land. Pump stations, which are spaced 20 to 100 miles apart along the underground pipelines, keep the petroleum products moving at around 5 miles per hour. At this rate, it takes 15 days to move a shipment of gasoline from Houston, Texas to New York City.

Americans use almost 17 million barrels of oil every day of the year. Texas produces more oil than any other state. The other top producing states are Alaska, California, Louisiana, and Oklahoma in that order. In all, 31 states produce petroleum.

The US has become increasingly dependent upon other countries for petroleum. In 1994 the US purchased 45% of its petroleum from other countries. Americans know this dependence can be dangerous. For example, the OPEC oil embargo was a political statement against the US due to its support of Israel in the Yom Kippur War. OPEC's oil embargo caused a drastic increase in the price of petroleum and made rationing an important part of the US economy in 1973. OPEC members try to set production levels for petroleum. As a rule, the less oil they produce, the higher the price of oil on the world market. The more oil they produce, the lower the price. However, OPEC countries do not always agree on the price of oil, where some want to produce less oil to raise prices, while others want to flood the market to benefit from immediate returns.

The US has also turned to non-Arab and non-OPEC countries for oil imports. Today, the US imports much of its oil from Canada and Mexico. This is beneficial because of the friendly relations the US has with its neighbors, and because the oil does not have to be shipped so far. Still, the amount of oil that the US can import from Canada and Mexico is limited. By law, Mexico can only export half the oil it produces to the US.

The usual basis for demand and supply forecasts is the World Energy Outlook (WEO) biannually prepared by the International Energy Agency (IEA). Table 1.3 shows 1973 and 2006 regional shares of crude oil production (IEA 2007).

1.2.2 Classification of Oils

The oil industry classifies "crude" by the location of its origin (e.g., "West Texas Intermediate, WTI" or "Brent") and often by its relative weight (API gravity), or viscosity ("light", "intermediate," or "heavy"), or sulfur content ("sweet" or "sour"). That is, refiners may refer to it as "sweet," which means the crude contains relatively little sulfur, or as "sour," which means it contains substantial amounts of sulfur and requires more refining in order to meet current product specifications.

Table 1.3 The 1973 and 2006 regional shares of crude oil production

Region	1973	2006
Middle East (%)	37.0	31.1
OECD (%)	23.6	23.2
Former USSR (%)	15.0	15.2
Africa (%)	10.0	12.1
Latin America (%)	8.6	9.0
Asia, excludes China (%)	3.2	4.5
China (%)	1.9	4.7
Non-OECD Europe (%)	0.7	0.2
Total (million tons)	2867	3936

There are different classification schemes, which are based on economic and/or geological criteria. There generally are two main oil classes: conventional oil and non-conventional oil. Conventional oil is oil that can be produced with current technology under present economic conditions. Table 1.4 shows the classification of oils.

Table 1.4 Classification of oils

	Viscosity of oil (measured in °API)
Conventional oil	
Crude oil	>17
Heavy oil	10–17
Deep sea oil above 500 m water depth	–
Non-conventional oil	
Extra-heavy oil below (including tar sands)	<10
Oil shale	–
Deep sea oil below 500 m water depth	–
Polar oil	–
Condensate	
Natural gas liquids (NGL)	–
Synthetic crude oil (SCO)	–
Bitumen from tar sands	–

1.2.3 Petroleum-Based Transportation Fuels

Refinery yield is the measure analysts use to assess how the breakdown of petroleum products coming from crude or unfinished oils varies over time. Refiners can adjust product yields in response to changing product prices and other market conditions. Table 1.5 shows minimum, average, and maximum refinery yields for selected products from crude oils. The amount of gasoline that can be distilled from crude depends on where the crude is obtained. Venezuelan crude yields little gasoline (about 5%), whereas Texas or Arabian crude yields about 30% gasoline. This is called “straight run” gasoline. Average gasoline yields at US refineries averaged 45.5% during 2007.

Crude oil is separated by boiling points into six main grades of hydrocarbons: refinery gas (used for refinery fuel), gasoline (naphthas), kerosene, light oils (diesel oil or diesel fuel) and heavy gas oils (fuel oil) and long residue. This initial separation is done by distillation. The first step in the refining of crude oil, whether in a simple or a complex refinery, is the separation of the crude oil into fractions (fractionation or distillation). These fractions are mixtures containing hydrocarbon compounds whose boiling points lie within a specified range.

Gasoline is a petroleum-derived liquid mixture, primarily used as fuel in internal combustion engines, specifically in spark-ignition engines. It consists mostly of aliphatic hydrocarbons, enhanced with iso-octane or the aromatic hydrocarbons toluene and benzene to increase its octane rating.

The density of gasoline is 0.71–0.77 kg/L. Gasoline is more volatile than diesel oil, Jet-A or kerosene, not only because of the base constituents, but also due to the additives that are put into it. The final control of volatility is often achieved by blending with butane. The desired volatility depends on the ambient temperature: in hotter climates, gasoline components of higher molecular weight and thus lower volatility are used. In cold climates, too little volatility results in cars failing to start. In hot climates, excessive volatility results in what is known as “vapor lock,” where combustion fails to occur because the liquid fuel has changed into a gaseous fuel in the fuel lines, rendering the fuel pump ineffective and starving the engine of fuel.

Table 1.5 Main crude oil refinery products (wt.%)

Product	Minimum	Maximum	Average
Liquefied petroleum gas	2	4	3
Gasoline	5	46	20
Jet fuel	6	12	10
Kerosene	7	13	10
No. 2 diesel fuel	16	30	23
Fuel oils	4	50	36
Heavy petroleum products	10	20	14

An important characteristic of gasoline is its octane rating, which is a measure of how resistant gasoline is to the abnormal combustion phenomenon known as predetonation (also known as knocking, pinging, spark knock, and other names). Octane rating is measured relative to a mixture of 2,2,4-trimethylpentane and *n*-heptane.

Diesel fuel is produced by distilling raw oil, which is extracted from bedrock. Diesel is a fossil fuel. Diesel fuel consists of hydrocarbons with between 9 and 27 carbon atoms in a chain as well as a smaller amount of sulfur, nitrogen, oxygen and metal compounds. It is a general property of hydrocarbons that the auto-ignition temperature is higher for more volatile hydrocarbons. The hydrocarbons present in diesel fuels include alkanes, naphthenes, olefins and aromatics. In addition other substances are added to improve the characteristics of diesel fuel. Its boiling point is between 445 and 640 K. A good diesel fuel is characterized by low sulfur and aromatic content, good ignition quality, the right cold weather properties, low content of pollutants and also the right density, viscosity and boiling point.

Diesel fuel comes in several different grades, depending upon its intended use. Like gasoline, diesel fuel is not a single substance, but a mixture of various petroleum-derived components, including paraffins, iso-paraffins, napthenes, olefins and aromatic hydrocarbons, each with their own physical and chemical properties.

Cetane number, volatility, and viscosity are the most important characteristics of diesel fuel. Cetane number (CN) is a measure of the ignition quality of the fuel. CN is based on two compounds, namely hexadecane with a CN of 100 and heptamethylnonane with a CN of 15. CN affects combustion roughness. Consumers often think the CN is similar to the octane number for gasoline, but that is not the case. Octane is a measure of a spark-ignition engine fuel's (gasoline) ability to resist engine knock (preignition from compression). Diesel cetane ratings work in the opposite direction. The higher the cetane rating, the more easily it ignites. Reaching desired cetane levels also limits the aromatic content of diesel fuel. Diesel fuel cetane ratings are calculated by calibrating a fuel to a mixture of reference fuels in a specially designed cooperative fuel research (CFR) engine. Various tests have been developed to calculate the CN from certain fuel properties. These tests usually involve some combination of fuel density and distillation properties. The two more commonly used CN estimate formulas are referred to as cetane indices to distinguish their results from the engine test. The most common cetane indices are ASTM D 976 and ASTM D 4737.

Unlike spark-ignition engines, the power and economy of diesel engines are comparatively insensitive to fuel volatility. There is some indirect impact in that less volatile fuels have higher heating values (HHV). Conversely, fuels with higher front-end volatility tend to improve starting and warm-up performance and reduce smoke. Ideal fuel volatility requirements will vary based on engine size and design, speed and load conditions, and atmospheric conditions. As an example, more volatile fuels may provide better performance for fluctuating loads and speeds such as those experienced by trucks and buses.

The viscosity of diesel fuel is an important property, which impacts the performance of fuel injection systems. Some injection pumps can experience excessive wear and power loss due to injector or pump leakage if viscosity is too low. If fuel

Table 1.6 Typical composition of gasoline

General name	Examples	Percentage
Aliphatic: straight chain	Heptane	15–25
Aliphatic: branched	Iso-octane	15–25
Aliphatic: cyclic	Cyclopentane	20–30
Aromatic	Ethyl benzene	20–30

viscosity is too high, it may cause too much pump resistance, filter damage, and adversely affect fuel spray patterns. High fuel viscosity can cause an injector spray pattern with poor fuel dispersion.

Gasoline is a petroleum-derived liquid mixture, primarily used as fuel in internal combustion engines, specifically in spark-ignition engines. In the Otto cycle engine a mixture of gasoline and air is compressed and is then ignited by a spark plug. The Otto cycle engine was invented by Dr. Nikolaus August Otto in 1876, also commonly known as a four-stroke cycle engine because of the four strokes involved in an internal combustion engine.

The important characteristics of gasoline are density, vapor pressure, distillation range, octane number, and chemical composition. To be attractive, a motor gasoline must have (a) desirable volatility, (b) anti-knock resistance (related to octane rating), (c) good fuel economy, (d) minimal deposition on engine component surfaces, and (e) complete combustion and low pollutant emissions.

An important characteristic of gasoline is its octane number or octane rating, which is a measure of how resistant gasoline is to the abnormal combustion phenomenon known as predetonation. Higher octane numbers are preferred in internal combustion engines. For gasoline production, aromatics, naphthenes and iso-alkanes are highly desirable, whereas olefins and *n*-paraffins are less desired. Typical composition of gasoline is given in Table 1.6.

1.3 Coal, Coal Liquefaction, and Refining

Coal continues to be the most widely used fossil fuel over the other energy sources because of its production in more than 50 countries, and because of the least effect it has on the price fluctuation. Coal will continue being an important resource in energy production with its long supply life of more than 200 years.

The first known and oldest fossil fuel is coal. Coal has played a key role as a primary energy source as well as a primary source of organic chemicals. It is a complex, heterogeneous combustible material, made up of portions that are either useful (carbon and hydrogen), or useless (diluents such as moisture, ash and oxygen, or contaminants such as sulfur and heavy metals). Coal can be defined as a sedimentary rock that burns. It was formed by the decomposition of plant matter, and it is a complex substance that can be found in many forms. Coal is divided into

Table 1.7 Chemical properties of typical coal samples

	Low-rank coal	High-volatile coal	High-rank coal
Carbon (%)	75.2	82.5	90.5
Hydrogen (%)	6.0	5.5	4.5
Oxygen (%)	17.0	9.6	2.6
Nitrogen (%)	1.2	1.7	1.9
Sulfur (%)	0.6	0.7	0.5
Moisture (%)	10.8	7.8	6.5
Calorific value (MJ/kg)	31.4	35.0	36.0

four classes: anthracite, bituminous, subbituminous, and lignite. Elemental analysis gives empirical formulas such as $C_{137}H_{97}O_9NS$ for bituminous coal and $C_{240}H_{90}O_4NS$ for high-grade anthracite.

Coal is formed from plant remains that have been compacted, hardened, chemically altered, and metamorphosed underground by heat and pressure over millions of years. When plants die in a low-oxygen swamp environment, instead of decaying by bacteria and oxidation, their organic matter is preserved. Over time, heat and pressure remove the water and transform the matter into coal. The first step in coal formation yields peat, compressed plant matter, which still contains leaves and twigs. The second step is the formation of brown coal or lignite. Lignite has already lost most of the original moisture, oxygen, and nitrogen. It is widely used as a heating fuel but is of little chemical interest. The third stage, bituminous coal, is also widely utilized as a fuel for heating. Bituminous is the most abundant form of coal and is the source of coke for smelting, coal tar, and many forms of chemically modified fuels. The chemical properties of typical coal samples are given in Table 1.7. Table 1.8 shows the global recoverable coal reserves (IEA 2007).

The worldwide coal production and consumption in 1998 were 4574 and 4548 million tons, respectively. The known global recoverable coal reserves in 1999 were 989 billion tons. Also, coal reserves are rather evenly spread around the globe: 25% are in the USA, 16% in Russia, and 11.5% in China. Although coal is much more abundant than oil or gas on a global scale, coalfields can easily get depleted on a regional scale.

Due to its abundance and wide distribution, coal accounts for 25% of the world's primary energy consumption and 37% of the energy consumed worldwide for electricity generation. For example, the known coal reserves in the world are enough for consumption for over 215 years, while the known oil reserves were only about 39 times the world's consumption and the known natural gas reserves were about 63 times the world's consumption level in 1998. With modern techniques coal can be mined, transported, and stored efficiently and cost effectively. International coal trade is growing steadily and the prices are kept low by the vigorous competition of supply. But coal's future commercial development depends critically on its environmental acceptability, and in particular on the success of the power generation industry in reducing polluting emissions.

Table 1.8 Global recoverable coal reserves

Country	Bituminous including anthracite	Subbituminous	Lignite
United States	115 891	101 021	33 082
China	62 200	33 700	18 600
India	82 396	–	2000
South Africa	49 520	–	–
Kazakhstan	31 100	–	3000
Brazil	–	11 929	–
Colombia	6267	381	–
Canada	3471	871	2236
Indonesia	790	1430	3150
Botswana	4300	–	–
Uzbekistan	1000	–	3000
Turkey	278	761	2650
Pakistan	–	2265	–
Thailand	–	–	1268
Chile	31	1150	–
Mexico	860	300	51
Peru	960	–	100
Kyrgyzstan	–	–	812
Japan	773	–	–
Korea (Democ. People's Rep.)	300	300	–
Zimbabwe	502	–	–
Venezuela	479	–	–
Philippines	–	232	100
Mozambique	212	–	–
Swaziland	208	–	–
Tanzania	200	–	–
Others	449	379	27

The major non-fuel use of coal is carbonization to make metallurgical coke. The production of activated carbon from coals has been of interest for years. Carbon in coal can be used as a source of specialty aromatic and aliphatic chemicals via processed including gasification, liquefaction, direct conversion, and co-production of chemicals, fuels and electricity.

Coal has played a key role as a primary source of organic chemicals as well as a primary energy source. Coal may become more important both as an energy source and as the source of carbon-based materials, especially aromatic chemicals in the twenty-first century (Schobert and Song 2002). New clean coal technologies can substantially improve efficiency and reduce emissions from power plants. In coal gasification the principal source of hydrogen is high-temperature steam. Steam is easily generated from inexpensive, abundant water. In coal liquefaction, on the other hand, hydrogen is usually added as molecular hydrogen H_2, which can be expensive to produce on a large scale.

The objective of coal gasification is to convert coal to a combustible gas, suitable for use as a fuel. The simplest way to obtain a combustible gas from coal is to carbonize the coal, that is, to heat it in the absence of air. The gaseous product from coal is called coal gas. Coal gas was first used in 1798 as a fuel for gas lamps that illuminated a factory owned by James Watt. The first company established in the US to manufacture and distribute coal gas was founded in Baltimore in 1817.

Coal consisting of carbon-based substances can be used as a source of specialty aromatic and aliphatic chemicals. There are four widespread processes that allow for making chemicals from coals: gasification, liquefaction, direct conversion, and co-production of chemicals and fuels along with electricity. Coal is gasified to produce synthesis gas (syngas) with a gasifier, which is then converted to paraffinic liquid fuels and chemicals by the Fischer–Tropsch synthesis. Liquid products from coal gasification mainly contain benzene, toluene, xylene (BTX), phenols, alkyl phenols, and cresol. Methanol can be made by using coal or syngas with hydrogen and carbon monoxide in a 2:1 ratio. Coal-derived methanol has many preferable properties such as being free of sulfur and other impurities. Syngas from coal can be reformed to hydrogen. Ammonium sulfate from coal tar by pyrolysis can be converted to ammonia. The humus substances can be recovered from brown coal by alkali extraction.

As early as 1800, coal gas or town gas was made by heating coal in the absence of air. Coal gas is rich in CH_4 and gives off up to 20.5 kJ per L of gas burned. Coal gas became so popular that most major cities and many small towns had a local gas house in which it was generated, and gas burners were adjusted to burn a fuel. An ideal coal gasification process would convert the coal completely to gas. Coal can be converted to water gas with steam (Demirbas 2007). The key chemical step for achieving this is the reaction:

$$C + H_2O \rightarrow CO + H_2 \tag{1.1}$$

which takes place at temperatures above 1075 K. Water gas burns to give CO_2 and H_2O, releasing roughly 11.2 kJ per L of gas consumed.

Water gas formed by the reaction of coal with oxygen and steam is a mixture of CO, CO_2, and H_2. The ratio of H_2 to CO can be increased by adding water to this mixture, to take advantage of a reaction known as the water–gas shift reaction (Demirbas 2007).

$$CO + H_2O \rightarrow CO_2 + H_2 \tag{1.2}$$

The concentration of CO_2 can be decreased by reacting the CO_2 with coal at high temperatures to form CO:

$$C + CO_2 \rightarrow 2CO \tag{1.3}$$

Water gas from which the CO_2 has been removed is called synthesis gas because it can be used as a starting material for a variety of organic and inorganic compounds. It can be used as the source of H_2 for the synthesis of methanol, for example.

$$CO + 2H_2 \rightarrow CH_3OH \tag{1.4}$$

Methanol can then be used as a starting material for the synthesis of alkenes, aromatic compounds, acetic acid, formaldehyde, and ethyl alcohol (ethanol). Synthesis gas can also be used to produce methane, or synthetic natural gas (Demirbas 2007).

The carbon monoxide/hydrogen mixture cannot be used as a direct replacement for methane. An additional process, methanation, is needed:

$$CO + 3H_2 \rightarrow CH_4 + H_2O \tag{1.5}$$

$$2CO + 2H_2 \rightarrow CH_4 + CO_2 \tag{1.6}$$

The methane produced in this way is sometimes called substitute natural gas or SNG. It would be easier, and possibly cheaper, to obtain SNG in a single processing step:

$$C + 2H_2 \rightarrow CH_4 \tag{1.7}$$

This process concept is called hydrogasification. It has not been used commercially, but has been the subject of much research and development effort. A promising new development in coal gasification is the integrated gasification combined-cycle plant (IGCC). It is being advertised by the electric power industry as the "cleanest method available for making electricity from coal."

The first step toward making liquid fuels from coal involves the manufacture of synthesis gas (CO and H_2) from coal. Coal can be gasified to produce synthesis gas (syngas), which can be converted to paraffinic liquid fuels and chemicals by the Fischer–Tropsch synthesis (FTS), which was developed in 1925 by Franz Fischer and Hans Tropsch to converted CO and H_2 at 1 atm and 575 K into liquid hydrocarbons using a Fe/Co catalyst. By 1941, Fischer–Tropsch plants produced 740 000 tons of petroleum products per year in Germany (Dry 1999). Fischer–Tropsch technology is based on a complex series of reactions that use H_2 to reduce CO to CH_2 groups linked to form long-chain hydrocarbons (Schulz 1999).

$$nCO + (n + m/2)\,H_2 \rightarrow C_nH_m + nH_2O \tag{1.8}$$

At the end of World War II, production of liquid fuels using Fischer–Tropsch technology was under study in most industrial nations. The low cost and high availability of crude oil, however, led to a decline in interest in liquid fuels made from coal.

Two basic approaches are used for coal liquefaction: indirect and direct. Indirect liquefaction involves the gasification of coal to produce a mixture of carbon monoxide and hydrogen (syngas). The syngas is then converted into a variety of liquid hydrocarbons using the FTS. Depending on the specific conditions (temperature, pressure, and catalyst) used for the FTS, many useful products can be made, including branched-chain paraffins (for high octane gasoline), kerosene and diesel fuel, fuel oil, and oxygen containing compounds that can in turn be converted to products such as synthetic detergents, synthetic margarine and cooking oil. Direct liquefaction is the reaction of coal with hydrogen (usually in the presence of some liquid solvent) to produce a synthetic crude oil, or syncrude. No intermediate gasification step is needed. Direct liquefaction is a very difficult process to carry out, involving temperatures over 675 K, hydrogen pressures of over 100 atm and an appropriate catalyst. The syncrude can be refined to produce gasoline with 75 to 80 octane rating, as well as diesel fuel and fuel oils. Today even modern versions of direct liquefaction, the result of four decades of research, are not competitive when crude oil sells for $25 per barrel. The economics of coal-derived liquids are generally thought to become attractive when the oil price exceeds $35 per barrel.

Another approach to liquid fuels is based on the reaction between CO and H_2 to form methanol (CH_3OH):

$$CO + 2H_2 \rightarrow CH_3OH \tag{1.9}$$

Methanol can be used directly as a transportation fuel, or it can be converted into gasoline with catalysts such as the ZSM-5 zeolite catalyst. Methanol can also be produced from syngas with hydrogen and carbon monoxide in a 2:1 ratio. Coal-derived methanol typically has low sulfur and few impurities.

1.4 Natural Gas

Natural gas was known in England as early as 1659, but it did not replace coal gas as an important source of energy in the world until after World War II. The usefulness of natural gas (NG) has been known for hundreds of years. For example, the Chinese used NG to heat water. In the early days, NG was used to light lamps on the street and in houses.

Natural gas is a mixture of lightweight alkanes. Natural gas contains methane (CH_4), ethane (C_2H_6), propane (C_3H_8), butane and iso-butane (C_4H_{10}), and pentanes (C_5H_{12}). The C_3, C_4, and C_5 hydrocarbons are removed before the gas is sold. The commercial natural gas delivered to the customer is therefore primarily a mixture of methane and ethane. The propane and butanes removed from natural gas are usually liquefied under pressure and sold as liquefied petroleum gases (LPG). Natural gas is found to consist mainly of the lower paraffins, with varying quantities of carbon dioxide, carbon monoxide, hydrogen, nitrogen and oxygen, in some cases also hydrogen sulfur and possibly ammonia. The chemical composition of NG is given in Table 1.9.

In recent years, NG has been the fastest growing primary energy source in the world, mainly because it is a cleaner fuel than oil or coal and not as controversial as nuclear power. NG combustion is clean and emits less CO_2 than all other petroleum derivatives, which makes it favorable in terms of the greenhouse effect. NG is used across all sectors, in varying amounts, including in industrial, residential, electric generation, commercial and transportation.

Table 1.9 Chemical composition of NG

Component	Typical analysis (% by volume)	Range (% by volume)
Methane	94.9	87.0–96.0
Ethane	2.5	1.8–5.1
Propane	0.2	0.1–1.5
i-Butane	0.03	0.01–0.3
n-Butane	0.03	0.01–0.3
i-Pentane	0.01	Trace to 0.14
n-Pentane	0.01	Trace to 0.14
Hexanes plus	0.01	Trace to 0.06
Nitrogen	1.6	1.3–5.6
Carbon dioxide	0.7	0.1–1.0
Oxygen	0.02	0.01–0.1
Hydrogen	Trace	Trace to 0.02

NG is found in many parts of the world, but the largest reserves are in the former Soviet Union and Iran. Since the 1970s, the global natural gas reserves have generally increased each year, with the reserves for various countries shown in Table 1.10 (Demirbas 2009).

Around the world, NG use is increasing for a variety of reasons including price, environmental concerns, fuel diversification and/or energy security issues, market deregulation, and overall economic growth. Figure 1.3 shows production and consumption trends of natural gas for the last few decades. In NG consumption, the US ranks first, the former USSR region ranks second, and Europe ranks third. The largest NG producer is Russia, which is also the largest supplier of NG to Western Europe. Asia and Oceania import NG to satisfy their demands. The other regions are relatively minor producers and consumers of gas.

Compared to oil, only a moderate amount of NG is traded on world markets. Due to its low density, NG is more expensive to transport than oil. For example, a section of pipe in oil service can hold 15 times more energy than when used to transport high-pressure NG. Hence, the gas pipelines needs to have a much larger diameter and/or fluid velocity for a given energy movement. In fact, pipeline transportation is not always feasible because of the growing geographic distance between gas reserves and markets. Many of the importing countries do not wish to solely rely on NG import due the potential political instabilities that may affect the long pipeline routes. The alternate transport routes are by ships or railcars. But, for economical transport, sufficient energy needs to be packaged in the containers, which is done by liquefaction. A full-liquefied natural gas (LNG) chain consists of a liquefaction plant, low-temperature pressurized transport ships, and a regasification terminal. World LNG trade is currently about 60 million metric tons per year, some 65% of which is imported by Japan.

Table 1.10 Global natural gas reserves by country

Country	Reserves (trillion cubic meter)	Percentage of world total
Russian Federation	48.1	33.0
Iran	23.0	15.8
Qatar	8.5	5.8
United Arab Emirates	6.0	4.1
Saudi Arabia	5.8	4.0
United States	4.7	3.3
Venezuela	4.0	2.8
Algeria	3.7	2.5
Nigeria	3.5	2.4
Iraq	3.1	2.1
Turkmenistan	2.9	2.0
Malaysia	2.3	1.6
Indonesia	2.0	1.4
Uzbekistan	1.9	1.3
Kazakhstan	1.8	1.3
Rest of World	23.8	16.5

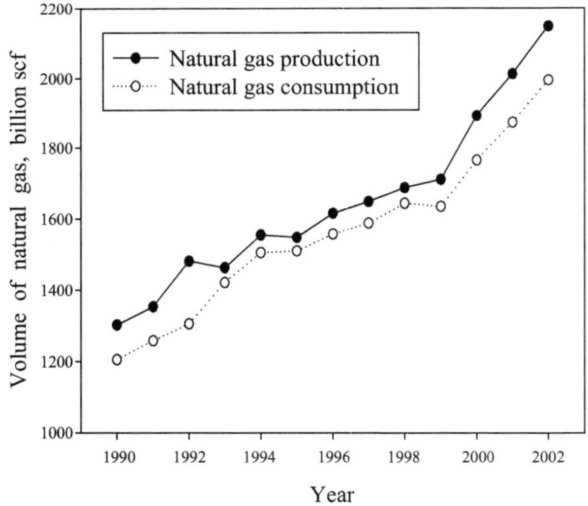

Fig. 1.3 Production and consumption trends of natural gas in the world

The generation of electricity is an important use of NG. However, the electricity from NG is generally more expensive than using coal because of increased fuel costs. NG can be used to generate electricity in a variety of ways: (1) conventional steam generation, similar to coal-fired power plants in which heating is used to generate steam which runs a turbine with an efficiency of 30–35%; (2) centralized gas turbine, in which hot gases from NG combustion are used to turn the turbine; and (3) combined-cycle unit, in which both steam and hot combustion gases are used to turn the turbine with an efficiency of 50–60%.

The use of NG in power production has increased due to the fact that NG is the cleanest burning alternative fossil fuel. Upon combustion, NG emits less CO_2 than oil or coal, virtually no sulfur dioxide, and only small amounts of nitrous oxides. CO_2 is a greenhouse gas, while sulfur and nitrous oxides produced by oil and coal combustion cause acid rain. Both the carbon and hydrogen in methane combine with oxygen when NG is burned, giving off heat, CO_2 and H_2O. Coal and oil contain proportionally more carbon than NG, hence emit more CO_2.

Concerns about acid rain, urban air pollution, and global warming are likely to increase NG use in the future. NG burns far more cleanly than gasoline or diesel, producing fewer nitrous oxides, unburned hydrocarbons and particulates.

$$CH_4 + 2O_2 \rightarrow CO_2 + 2H_2O \quad (1.10)$$

1.00 g 2.75 g

$$2C_4H_{10} + 13O_2 \rightarrow 8CO_2 + 10H_2O \quad (1.11)$$

1.00 g 3.03 g

$$C + O_2 \rightarrow CO_2 \quad (1.12)$$

1.00 g 3.67 g

From Eqs. 1.10–1.12, among the fossil fuels, natural gas is the least responsible for CO_2 emissions. Liquefied petroleum gas (LPG) causes higher levels of CO_2 than natural gas (Eq. 1.11). The highest amount of CO_2 occurs according to Eq. 1.12. Thus impact of the fossil fuel increases with an increase in its carbon number (Demirbas 2005). The gases (they consist of three or more atoms like CO_2 and CH_4) with higher heat capacities than those of O_2 and N_2 contribute to the greenhouse effect.

Since NG vehicles require large storage tanks, the main market may be for buses that are used within cities. Another use that may be developed is the use of fuel cells for stationary and transportation application. The energy for fuel cells comes from hydrogen, which can be made from NG. Fuel cells eliminate the need for turbines or generators, and can operate at efficiencies as high as 60%. In addition, fuel cells can also operate at low temperatures, reducing the emission of acid rain-causing nitrous oxides, which are formed during high-temperature combustion of any fuel.

Natural gas (methane) can be obtained from gas hydrates. Gas hydrates are also called methane hydrates or chemically clathrates. Gas hydrates are potentially one of the most important energy resources for the future. Methane gas hydrates in particular are increasingly considered a potential energy resource. Methane gas hydrates are crystalline solids formed by the combination of methane and water at low temperatures and high pressures. Gas hydrates have an ice-like crystalline lattice of water molecules with methane molecules trapped inside. Enormous reserves of hydrates can be found under continental shelves and on land under permafrost. The amount of organic carbon in gas hydrates is estimated to be twice that in all

Table 1.11 Worldwide amount of organic carbon sources

Source of organic carbon	Amount (gigaton)
Gas hydrates (onshore and offshore)	10 000–11 000
Recoverable and non-recoverable fossil fuels (oil, coal, gas)	5000
Soil	1400
Dissolved organic matter	980
Land biota	880
Peat	500
Others	70

other fossil fuels combined. However, due to the solid form of the gas hydrates, conventional gas and oil recovery techniques are not suitable. Table 1.11 shows the worldwide amount of organic carbon sources (Hacisalihoglu et al. 2008). The recovery of methane generally involves dissociating or melting in situ gas hydrates by heating the reservoir above the temperature of hydrate formation, or decreasing the reservoir pressure below hydrate equilibrium (Lee and Holder 2001).

The difficulty with recovering this source of energy is that the fuel is in solid form and is not amenable to conventional gas and oil recovery techniques (Lee and Holder 2001). Proposed methods of gas recovery from hydrates generally deal with dissociating or melting in situ gas hydrates by heating the reservoir beyond the temperature of hydrate formation, or decreasing the reservoir pressure below hydrate equilibrium. Models have been developed to evaluate natural gas production from hydrates by both depressurization and heating methods.

There are three methods to obtain methane from gas hydrates: (1) the depressurization method, (2) the thermal stimulation method, and (3) the chemical inhibition method. The thermal stimulation and chemical inhibitor injection methods are both costly procedures, whereas the depressurization method may prove useful when applied to more than one production.

1.5 Oil Shale and Tar Sands

Oil shale or shale oil is a sedimentary rock that contains the solid hydrocarbon wax kerogen in tightly packed limy mud and clay. Oil shale is an inorganic, non-porous rock that contains some organic material in the form of kerogen. In some respects, oil shale is similar to the source rock that produced petroleum. Oil shale contains more kerogen than oil source rock, but less than coal. The composition of the oil derived from oil shale is much more similar to the composition of petroleum than to coal. Some oil shales can be ignited, like coal, and they burn with a very sooty, smoky flame much like a coal of very high-volatile matter content. However, oil shales are not considered for use as solid fuels. Their principal application is in the conversion to liquid fuels.

Two thirds of the world's oil shale reserves are located in the US. The largest known reserves of hydrocarbons of any kind are the Green River shale deposits in Wyoming, Colorado, and Utah. These reserves are estimated to be 270 billion tons. At 20 gallons per ton of shale, this translates into 130 billion barrels of oil. This is five times as much as the proven reserves of petroleum in the US. However, no commercial production of fuels from oil shale exists today, so their economic recoverability is not well known.

The kerogen may be decomposed at elevated temperatures (723 K), resulting in an oil suitable for refinery processing (Dorf 1977). The oil shale layer is not hot enough to complete the oil generation. For the final step the kerogen must be heated up to 775 K and molecularly combine with additional hydrogen to complete the oil formation. This final process must be performed in the refinery and needs huge amounts of energy which otherwise were provided by the geological environment during oil formation (Demirbas 2000). The kerogen is still in the source rock and could not accumulate in oil fields. Typically, the ratio of kerogen to waste material is very low, making the mining of oil shales unattractive. Hence, due to a combination of environmental and economic concerns, it is very unlikely that oil shale mining will ever be performed at a large scale, though at some places it is has been utilized in small quantities. However, the shale oil reserves in the world are greater than crude oil or natural gas, as shown in Fig. 1.1.

The industrial processing of oil shale began a century ago. During the 1920s, oil shale was an economically important energy source. During the energy crisis of the 1970s several major oil companies made massive investments both in research and development and in possible commercial use of oil shale. However, several technical problems must be overcome for successful large-scale use of oil shale in the future. The composition of oil from oil shale is sufficiently different from that of petroleum that liquids derived from oil shale cannot be used as direct substitutes for petroleum. Oil derived from oil shale has less carbon and hydrogen and more nitrogen and oxygen than petroleum. Petroleum refinery operations would have to be modified to accommodate oil shale liquids as a feedstock. Hydrogen must be added to the oil during processing. When oil shale is retorted, the inorganic portion of the shale expands considerably. The liquid fuels derived from oil shale can become important alternatives to petroleum.

Tar sands are grains of sand or, in some cases, porous carbonate rocks that are intimately mixed with a very heavy, asphalt-like crude oil called bitumen. The bitumen is much too viscous to be recovered by traditional petroleum recovery techniques. Tar sands contain about 10–15% bitumen, the remainder being sand or other inorganic materials.

The estimated global resources of tar sands are about three times the known petroleum reserves. The world's largest deposit of tar sands is near Athabasca, in Alberta, Canada. Other large deposits exist in the former Soviet Union and in Venezuela. In the US, small deposits of tar sands are found in Utah.

Tar sands are oil traps not deep enough in the Earth to allow for geological conversion into the conventional oil. This oil was not heated enough to complete the process of molecular breakage to reduce the viscosity. The oil has the characteristics of bitumen and is mixed with large amounts of sand due to the proximity to the Earth surface. The tar sand is mined, flooded with water in order to separate the heavier sand, and then processed in special refineries to reduce its high sulfur content (the original oil usually has 3–5% sulfur) and other components. This process needs huge amounts of energy and water. Only oil deposits in deep layers below 75 m are mined in situ (COSO 2007).

If tar sand is heated to about 355 K, by injecting steam into the deposit in a manner analogous to that of enhanced oil recovery, the elevated temperature causes a decrease in the viscosity of the bitumen just enough to allow for its pumping to the surface. Alternatively, it is sometimes easier to mine the tar sand as a solid material. If the bitumen is heated up above 775 K this converts about 70% of it to a synthetic crude oil. Distilling this oil gives good yields of kerosene and other liquid products in the middle distillate range.

1.6 Natural Bitumen and Extra-Heavy Oil

Natural bitumen is the oil contained in clastic and carbonate reservoir rocks, most frequently in small deposits at, or near, the Earth's surface. Clastic rocks are composed of fragments of preexisting rock. These rocks are commonly referred to as tar sands or oil sands and have been mined since antiquity for use as paving. Natural bitumen is found in 183 identified deposits in 21 countries. Natural bitumen (often called tar sands or oil sands) and heavy oil differ from light oils by their high viscosity at reservoir temperatures, high density (low API gravity), and significant contents of nitrogen, oxygen, and sulfur compounds and heavy-metal contaminants.

Historically, heavy oil was found incidentally during the search for light oil and was produced by conventional methods when economically feasible. However, to sustain commercial well production rates, heavy and extra-heavy oil production almost always requires measures to reduce oil viscosity and to introduce energy into the reservoir. Extra-heavy oil commonly requires the addition of diluents (gas condensate, natural gas liquids, or light crude) to enable the oil to be transported by pipeline. Extra-heavy oil must also be chemically upgraded to reduce density and remove contaminants before it can be used as refinery feedstock.

Extra-heavy oil and natural bitumen represent crude oils, which have been severely degraded by microbial action, as evidenced by the paucity of low molecular weight saturated hydrocarbons. Extra-heavy oil is recorded in 91 deposits. Some of these represent separate reservoirs in a single field, of which some are producing and others abandoned.

The recoverable volumes of oil contained in deposits of extra-heavy oil and natural bitumen are immense. If oil, in place for conventional oil, is estimated to be about three times the original reserves, then the remaining oil, in place after the deduction of cumulative production, is about 4925 billion barrels. Figure 1.4 provides a comparison of distribution of the world's known recoverable oil resources and reserves by type. Table 1.12 shows comparative chemical analyses of world oils and natural bitumen.

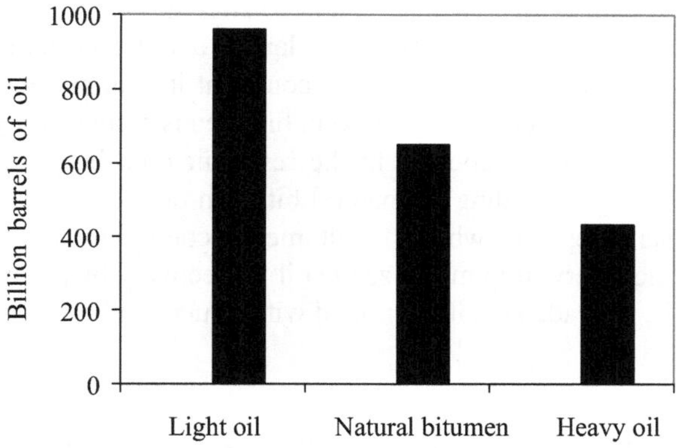

Fig. 1.4 Distribution of the world's known recoverable oil resources and reserves by type

Two fundamental upgrading processes are presently employed to prepare heavy oil and natural bitumen for transportation and refining to finished products. These processes are carbon rejection and hydrogen addition. Each process improves the hydrogen-to-carbon ratio but by following different paths. Carbon rejection yields a large quantity of low heating value gas at the expense of produced liquids, a large amount of petroleum coke, and therefore moderate conversion at low pressure. The yield of upgraded oil (synthetic crude oil) from the natural bitumen varies with the technology employed, the consumption of product for fuel in the upgrader, the extent of natural gas liquids recovery, and the degree of residue upgrading.

Recent estimates of world recoverable oil resources are consistent with an estimate of oil in place of conventional remaining oil reservoirs which are on the order of 4925 billion barrels, excluding extra-heavy oil and natural bitumen. The densities and viscosities are directly related to their origin as the degraded residues of con-

Table 1.12 Comparative chemical analyses of world oils and natural bitumen

	Conventional oil	Medium oil	Heavy oil	Extra-heavy oil	Natural bitumen
Coke in crude oil (wt.%)	10.4	17.6	21.8	28.1	–
Gasoline yields (vol.%)	9.2	2.8	2.0	1.3	1.4
Gas oil yield (vol.%)	17.4	21.9	15.9	16.9	7.2
Residuum yield (vol.%)	21.9	39.5	52.6	62.6	18.1
Crude oil density (kg/L)	0.836	0.920	0.958	1.018	1.041
Crude oil gravity (°API)	38.1	22.3	16.3	7.5	5.0
Crude oil viscosity (cP at 411 K)	9	63	593	7936	292 991
Resins (wt.%)	6.1	19.3	24.2	21.2	25.2
Asphaltenes (wt.%)	2.1	6.6	12.4	13.2	30.6
Sulfur (wt.%)	0.4	1.5	2.9	4.9	3.3

ventional oils. This has resulted in a large increase in content of polar molecules, of which the asphaltenes are the largest and most troublesome.

Natural bitumen is so viscous that it is immobile in the reservoir. For oil sand deposits less than 225 ft deep, bitumen is recovered by mining the sands, then separating the bitumen from the reservoir rock by processing it with hot water, and finally upgrading the natural bitumen onsite to a synthetic crude oil. In deeper oil sand deposits, where the bitumen is commonly less viscous, steam is injected into the reservoir to mobilize the oil for recovery by production wells. The product may be upgraded onsite or mixed with diluent and transported to an upgrading facility.

1.7 Introduction to Renewable and Biorenewable Sources

The major sources of alternative energy are biorenewables, hydro, solar, wind, geothermal and other energies, each of them having their own advantages and disadvantages, including political, economical and practical issues.

The term *renewable energy* refers to primary energies that are regarded as inexhaustible in terms of human (time) dimensions. They are continuously generated by the energy sources of solar energy, geothermal energy, and tidal energy. The energy produced within the Sun is responsible for a multitude of other renewable energies (such as wind and hydropower) as well as renewable energy carriers (such as solid or liquid biofuels). Main renewable energy sources are biomass, hydropower, geothermal, solar, wind, and marine energies. The renewables are the primary, domestic, and clean or inexhaustible energy resources (Kaltschmitt et al. 2007).

Renewable energy technologies produce marketable energy by converting natural phenomena into useful forms of energy. These technologies use the Sun's energy and its direct and indirect effects on the Earth (solar radiation, wind, falling water and various plants, i. e., biomass), gravitational forces (tides), and the heat of the Earth's core (geothermal) as the resources from which energy is produced (Kalogirou 2004). The renewable energies give rise to a multitude of very different energy flows and carriers due to various energy conversion processes occurring in nature.

Renewable energy is a promising alternative solution because it is clean and environmentally safe. They also produce lower or negligible levels of greenhouse gases and other pollutants when compared with the fossil energy sources they replace.

1.7.1 Non-Combustible Renewable Energy Sources

Non-combustible renewable energy sources are hydro, geothermal, wind, solar, wave, tidal, and ocean thermal energy sources. Non-combustible renewable energies have been the primary energy source in the history of the human race.

The water in rivers and streams can be captured and turned into hydropower, also called hydroelectric power. Large-scale hydropower provides about one-quarter of the world's total electricity supply, virtually all of Norway's electricity, and more than 40% of the electricity used in developing countries. The technically usable global potential of large-scale hydro is estimated to be over 2200 GW, of which only about 25% is currently exploited.

There are two small-scale hydropower systems: micro hydropower systems (MHP) with capacities below 100 kW, and small hydropower systems (SHP) with capacity between 101 kW and 1 MW. Large-scale hydropower supplies 20% of global electricity. In the developing countries, considerable potential still exists, but large hydropower projects may face financial, environmental, and social constraints (UNDP 2000). There is no universal consensus on the definition of small hydropower. Some countries of the European Union such as Portugal, Spain, Ireland, Greece, and Belgium accept 10 MW as the upper limit for installed capacity. In Italy the limit is 3 MW, in France 8 MW, in the UK 5 MW, in Canada 20–25 MW, in the US 30 MW; however, a value of up to 10 MW total capacities is becoming generally accepted as small hydropower in the rest of the world. If total installed capacity of any hydropower system is bigger than 10 MW, it is generally accepted as a large hydropower system (Demirbas 2006). Small-scale hydropower can be further subdivided into minihydro usually defined as <500 kW and microhydro is <100 kW. Small hydropower is one of the most valuable energies to be offered to rural communities' electrification and its growth is to decrease the gap of decentralized production for private sector and municipal activity production. Small-scale hydropower systems supply energy from flowing or running water and convert it to electrical energy, where the potential for small hydropower systems depends on the availability of water flow where the resource exists. If a well-designed small hydropower system is established anywhere, it can fit in with its surroundings and will have minimal negative impacts on the environment. Small hydropower systems allow for self-sufficiency by using the scarce natural water resources. These systems supply low-cost energy production that is applied in many developing countries throughout the world.

Dams are individually unique structures and dam construction represents the biggest structures in terms of basic infrastructure within all nations. Today, nearly 500 000 square kilometers of land are inundated by reservoirs in the world, capable of storing 6000 km^3 of water.

The efficiency of small hydropower depends mainly on the performance of the turbine. Today, generators commonly have efficiency rates of 98–99%. As a general rule, larger and newer plants have higher efficiencies of up to 90%. Efficiency can be as low as 60% for old and smaller plants. Hydropower is the most efficient way to generate electricity. Modern hydroturbines can convert as much as 90% of the available energy into electricity. The best fossil fuel plants are only about 50% efficient.

Geothermal energy can be utilized in various forms such as electricity generation, direct use, space heating, heat pumps, greenhouse heating, and industrial. The electricity generation is improving faster in geothermal energy-rich countries. As an energy source, geothermal energy has come of age. The utilization has increased rapidly during the last three decades.

Geothermal energy for electricity generation has been produced commercially since 1913, and for four decades on the scale of hundreds of megawatts both for

Table 1.13 World's top countries using geothermal energy in direct uses

Country	Installed (MWt)	Production (GWh/yr)
China	2282	10 531
Japan	1167	7482
United States	3766	5640
Iceland	1469	5603
Turkey	820	4377
New Zealand	308	1967
Georgia	250	1752
Russia	308	1707
France	326	1360
Sweden	377	1147
Hungary	473	1135
Mexico	164	1089
Italy	326	1048
Romania	152	797
Switzerland	547	663

electricity generation and direct use. In 2000, geothermal resources have been identified in over 80 countries and there are quantified records of geothermal utilization in 58 countries in the world (Fridleifsson 2001). The world's top countries using geothermal energy directly are given in Table 1.13.

Geothermal energy is clean, cheap and renewable, and can be utilized in various forms such as space heating and domestic hot water supply, CO_2 and dry-ice production process, heat pumps, greenhouse heating, swimming and balneology (therapeutic baths), industrial processes and electricity generation. The main types of direct use are bathing, swimming and balneology (42%), space heating (35%), greenhouses (9%), fish farming (6%), and industry (6%) (Fridleifsson 2001). Electricity is produced with geothermal steam in 21 countries spread over all continents. Low-temperature geothermal energy is exploited in many countries to generate heat, with an estimated capacity of about 10 000 MW.

The direct application of geothermal, including space heating and cooling, industry, greenhouses, fish farming, etc., uses mostly existing technology and straightforward engineering. The technology, reliability, economics, and environmental acceptability of direct use of geothermal energy have been demonstrated throughout the world.

Wind power has long been used for grain-milling and water-pumping applications. Significant technical progress since the 1980s, driven by advances in aerodynamics, materials, design, controls, and computing power, has led to economically competitive electrical energy production from wind turbines. Technology development, favorable economic incentives (due to its early development status and environmental benefits), and increasing costs of power from traditional fossil sources have led to significant worldwide sales growth since the early 1980s (Kutz 2007).

Wind energy is renewable, and its generation is characterized by large variations in production. The wind energy sector is one of the fastest growing energy sectors in the world. The global interest in wind power engineering entered during industrial development. From 1991 until the end of 2002, the global installed capacity increased from about 2 GW to over 31 GW, with an average annual growth rate of about 26%. During this period, both prices of wind turbines and cost of wind-generated electricity have reduced. Wind power would need to be produced 1% more to compensate for the losses of hydropower production. Wind power production, on an hourly level for 1–2 days ahead, is more difficult to predict than other production forms (Demirbas 2006).

Wind energy is a significant resource; it is safe, clean, and abundant. Wind energy is an indigenous supply permanently available in virtually every nation in the world. Wind energy is abundant, renewable, widely distributed, clean, and mitigates the greenhouse effect if it is used to replace fossil fuel-derived electricity. Wind energy has limitations based on geography and meteorology, plus there may be political or environmental problems (e. g., dead birds) with installing turbines (Garg and Datta 1998). On the other hand, wind can cause air pollution by degradation and distribution of pieces of pollutants such as waste paper, straw, etc.

The detailed analysis of the technical, economic and regulatory issues of wind power is discussed in the European Wind Energy Association (EWEA) report, "Large-Scale Integration of Wind Energy in the European Power Supply: Analysis, Issues and Recommendations" published in December 2005. In 2005, worldwide capacity of wind-powered generators was 58 982 megawatts; although wind power currently produces less than 1% of worldwide electricity use, it accounts for 23% of electricity use in Denmark, 4.3% in Germany, and approximately 8% in Spain. Globally, wind power generation more than quadrupled between 1999 and 2005 according to the EWEA (2005).

Figure 1.5 shows the growth scenarios for global installed wind power (IEA 2006). In 2004, the International Energy Agency (IEA) Reference Scenario projections for wind energy were 66 GW in 2010, 131 GW in 2020, and 170 GW in 2030. Recently, the IEA Reference Scenario projections for wind energy were updated to 75 GW in 2010, 145 GW in 2020, and 202 GW in 2030. The IEA advanced strong growth scenario projected a wind energy market of 82 GW in 2010, 165 GW in 2020, and 250 GW in 2030.

Solar energy is defined as radiant energy transmitted by the Sun and intercepted by Earth. It is transmitted through space to Earth by electromagnetic radiation with wavelengths ranging between 0.20 and 15 μm. The availability of solar flux for terrestrial applications varies with season, time of day, location, and collecting surface orientation. Here, we shall treat these matters analytically (Kutz 2007).

Energy comes from processes called solar heating (SH), solar home heating (SHH), solar dryer (SD), solar cooker (SC), solar water heating (SWH), solar photovoltaic (SPV, converting sunlight directly into electricity), and solar thermal electric power (STEP, when the Sun's energy is concentrated to heat water and produce steam, which is used to produce electricity). The major component of any

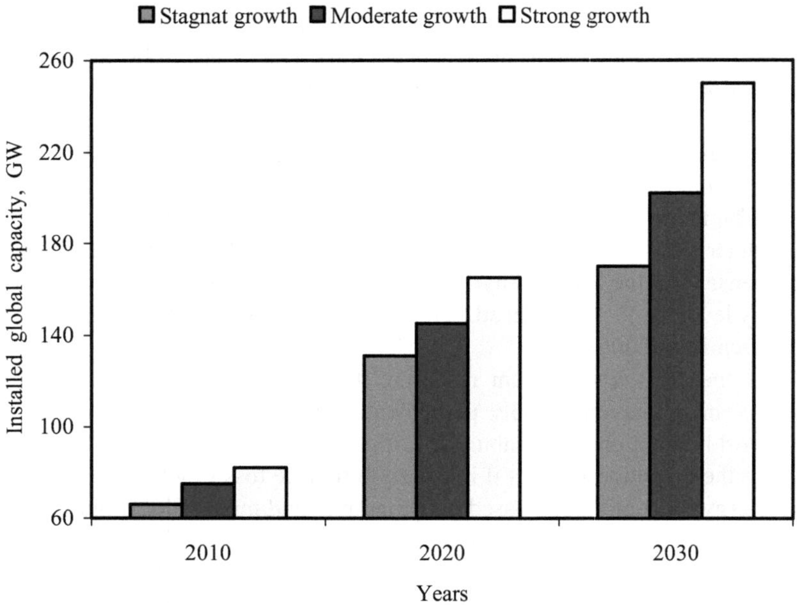

Fig. 1.5 Scenarios for global installed wind power

solar system is the solar collector. Solar energy collectors are a special kind of heat exchanger that transforms solar radiation energy to internal energy of the transport medium.

Solar dryers are used for drying fruits and spices. The three most popular types of SD are box type, cabinet type, and tunnel type. Box type uses direct heat for dehydration. In cabinet type dryers, air heated by the collector dehydrates the food product, whereas in tunnel type forced air circulation is used to distribute heat for dehydration. Cabinet and tunnel type dryers yield a high quality of dried products but they are very bulky and costly compared to the box type dryers. Of about 800 dryers disseminated so far, 760 are of the box type (Pokharel 2003).

One of the most abundant energy resources on the surface of the Earth is sunlight. Today, solar energy has a tiny contribution in the world total primary energy supply of less than 1%. Photovoltaic (PV) systems, other than solar home heating systems, are used for communication, water pumping for drinking and irrigation, and electricity generation. The total installed capacity of such systems is estimated at about 1000 kW. A solar home heating system is a solar PV system with a maximum capacity of 40 W. These systems are installed and managed by a household or a small community (Garg and Datta 1998). A historical introduction into the uses of solar energy, followed by a description of the various types of collectors including flat-plate, compound parabolic, evacuated tube, parabolic trough, Fresnel lens, parabolic dish and heliostat field collectors, is given by Kalogirou (2004).

Solar thermal electricity systems utilize solar radiation to generate electricity by the photothermal conversion method. Solar-based chemical processes utilize solar energy to drive chemical changes. These processes offset energy that would otherwise be required from an alternate source and can serve as a method of converting solar energy into a storable and transportable fuel. Solar chemical reactions are diverse but can generically be described as either thermochemical or photochemical. A more conventional approach uses process heat from a solar concentrator to drive the steam reformation of natural gas thereby increasing the overall hydrogen yield. Concentrators can be used in the gasification of feedstocks such as coal, woody biomass, municipal solid waste and agricultural residues. The resulting hydrocarbons can be used to synthesize so-called sunfuels. Concentrating solar technologies can also be used in the production of industrial chemicals. Photoelectrochemical cells (PECs) consist of a semiconductor, typically titanium dioxide or related titanates, immersed in an electrolyte. A photogalvanic device is a type of battery in which the cell solution forms energy-rich chemical intermediates when illuminated. These chemical intermediates then react at the electrodes to produce an electric potential.

Wave energy, tidal energy, and ocean thermal energy conversion (OTEC) are the other non-combustible renewable energy sources (RES). Water energy sources are hydro, tidal and wave technologies. Marine energy sources include current, tidal, OTEC and wave technologies. The world wave resource is between 200 and 5000 GW, mostly found in offshore locations (Garg and Datta 1998). Wave energy converters fixed to the shoreline are likely to be the first to be fully developed and deployed, but waves are typically two to three times more powerful in deep offshore waters than at the shoreline. Wave energy can be harnessed in coastal areas, close to the shore. The first patent for a wave energy device was filed in Paris in 1799, and by 1973 there were 340 British patents for wave energy devices. By comparison to wind and PV, wave energy and tidal stream are very much in their infancy. Currently, around 1 MW of wave energy devices are installed worldwide, mainly from demonstration projects.

The OTEC is an energy technology that converts solar radiation to electric power. OTEC systems use the ocean's natural thermal gradient to drive a power-producing cycle. As long as the temperature between the warm surface water and the cold deep water differs by about 20 K, an OTEC system can produce a significant amount of power. The oceans are thus a vast renewable resource, with the potential to help us produce billions of watts of electric power.

1.7.2 Combustible Renewable Energy Sources

Combustible renewable or biorenewable materials such as lignocellulosic materials, crops, grasses, animal wastes and biogas, are combustible renewable energy sources. Biomass resources include various natural and derived materials, such as woody and herbaceous species, wood wastes, bagasse, agricultural and industrial residues, waste paper, municipal solid waste, sawdust, biosolids, grass, waste from food processing, animal wastes, aquatic plants and algae, etc. Biomass is a solar energy carrier. Solar energy is stored in biomass as chemical energy via photosynthesis during the growth of plants and trees.

Table 1.14 Main biorenewable feedstocks

Biorenewable feedstocks	Description
Sucrose-containing crops	Sugar cane, sugar beet, beet molasses, sweet sorghum
Starchy crops	Corn, potato, starchy food wastes
Lignocellulosic materials	Corn stover, wheat straw, perennial crops
Vegetable oils	Produced from soybeans, canola, palm
Animal fats	Tallow, lard, fish oil
Recycled products	Yellow grease, brown grease, cooking oil
Pyrolysis oil	Wood residues, waste biomass, organic solid wastes

The replacement of petroleum with biorenewables for fuels and chemicals requires the identification of feedstock, intermediate, and product species that replace their fossil counterparts. Main biorenewable feedstocks are presented in Table 1.14.

Biofuels originate from plant oils, sugar beets, cereals, organic waste and the processing of biomass. Biological feedstocks that contain appreciable amounts of sugar – or materials that can be converted into sugar, such as starch or cellulose – can be fermented to produce ethanol to be used in gasoline engines. Ethanol feedstocks can be conveniently classified into three types: (1) sucrose-containing feedstocks, (2) starchy materials, and (3) lignocellulosic biomass. Feedstock for bioethanol is essentially comprised of sugar cane and sugar beet. The two are produced in geographically distinct regions. Sugar cane is grown in tropical and subtropical countries, while sugar beet is only grown in temperate climate countries.

Today the production cost of bioethanol from lignocellulosic materials is still too high, which is the major reason why bioethanol has not made its breakthrough yet. When producing bioethanol from corn or sugar cane the raw material constitutes about 40–70% of the production cost.

In European countries, beet molasses are the most utilized sucrose-containing feedstock. Sugar beet crops are grown in most of the EU-25 countries, and yield substantially more ethanol per hectare than wheat. The advantages with sugar beet are a lower cycle of crop production, higher yield, and high tolerance of a wide range of climatic variations, low water and fertilizer requirement. Sweet sorghum is one of the most drought resistant agricultural crops as it has the capability to remain dormant during the driest periods.

The conversion of sucrose into ethanol is easier compared to starchy materials and lignocellulosic biomass because previous hydrolysis of the feedstock is not required since this disaccharide can be broken down by the yeast cells; in addition, the conditioning of the cane juice or molasses favors the hydrolysis of sucrose.

Another type of feedstock, which can be used for bioethanol production, are starch-based materials. Starch is a biopolymer and is defined as a homopolymer, consisting of only one monomer, D-glucose. To produce bioethanol from starch it is necessary to break down the chains of this carbohydrate for obtaining glucose syrup, which can be converted into ethanol by yeasts. The single greatest cost in the production of bioethanol from corn, and the cost with the greatest variability, is the cost of the corn. Ethanol (ethyl alcohol) is the most important of biofuels. It can be used as a fuel, mainly as a transportation biofuel alternative to gasoline, and is widely used in cars in Brazil. Because it is easy to manufacture and process and can be made from very common crops such as sugar cane and corn, it is an increasingly common alternative to gasoline in some parts of the world. Anhydrous ethanol (ethanol with less than 1% water) can be blended with gasoline in varying quantities up to pure ethanol (E100), and most gasoline engines will operate well with mixtures of 10% ethanol (E10).

Lignocellulosic biomass, such as agricultural residues (corn stover and wheat straw), wood and energy crops, are attractive materials for ethanol fuel production since they are the most abundant reproducible resources on Earth. Lignocellulosic perennial crops (e. g., short rotation coppices and grasses) are promising feedstock because of high yields, low costs, good suitability for low-quality land (which is more easily available for energy crops), and few environmental impacts.

References

COSO (Crude Oil Supply Outlook). 2007. Report to the Energy Watch Group EWG-Series No. 3/2007. http://www.energywatchgroup.org/fileadmin/global/pdf/EWG-Oilreport-10-2007.pdf. Accessed 2009.

Demirbas, A. 2000. Carbonization and characterization of Turkish oil shales. Energy Sources 22:675–682.

Demirbas, A. 2005. Potential applications of renewable energy sources, biomass combustion problems in boiler power systems and combustion related environmental issues. Progress Energy Combus Sci 31:171–192.

Demirbas, A. 2006. Energy priorities and new energy strategies. Energy Edu Sci Technol 16:53–109.

Demirbas, A. 2007. Utilization of coals as sources of chemicals. Energy Sources 29:677–684.

Demirbas, A. 2009. Biofuels: securing the planet's future energy needs. Springer, London.

Dorf, R.C. 1977. Energy resources and policy. Addison-Wesley, Los Angeles, CA.

Dry, M.E. 1999. Fischer–Tropsch reactions and the environment. Appl Catal A General 189:185–190.

EWEA (European Wind Energy Association). 2005. Report: large scale integration of wind energy in the European power supply: analysis, issues and recommendations. EWEA, Brussels, Belgium.

Fridleifsson, I.B. 2001. Geothermal energy for the benefit of the people. Renew Sustain Energy Rev 5:299–312.

Garg, H.P., Datta, G. 1998. Global status on renewable energy, in Solar Energy Heating and Cooling Methods in Building, International Workshop, Iran University of Science and Technology, 19–20 May 1998.

Hacisalihoglu, B., Demirbas, A.H., Hacisalihoglu, S. 2008. Hydrogen from gas hydrate and hydrogen sulfide in the Black Sea. Energy Edu Sci Technol 21:109–115.

IEA (International Energy Agency). 2006. Reference scenario projections. IEA, Paris.

IEA, 2007. The International Energy Agency (IEA). Key world energy statistics. IEA, Paris.

http://www.iea.org/Textbase/nppdf/free/2007/key_stats_2007.pdf. Accessed 2009.

Kalogirou, A.S. 2004b. Solar thermal collectors and applications. Progress Energy Combus Sci 30:231–295.

Kaltschmitt, M., Streicher, W., Wiese, A. (Eds.) 2007. Renewable energy: technology economics and environment. Springer, Berlin Heidelberg New York.

Kutz, M. (Ed.) 2007. Environmentally conscious alternative energy production. Wiley, Hoboken, NJ.

Lee, S.-Y., Holder, G.D. 2001. Methane hydrates potential as a future energy source. Fuel Proc Technol 71:181–186.

Pokharel, S. 2003. Promotional issues on alternative energy technologies in Nepal. Energy Policy 31:307–318.

Schobert, H.H., Song, C. 2002. Chemicals and materials from coal in the 21st century. Fuel 81:15–32.

Schulz, H. 1999. Short history and present trends of FT synthesis. Appl Catal A General 186:1–16.

UNDP (United Nations Development Programme). 2000. World energy assessment 2000: energy and the challenge of sustainability. UNDP, New York.

Chapter 2
Fuels from Biomass

2.1 Introduction

Due to increasing environmental concerns especially related with the use of fossil fuels, new solutions to limit the greenhouse gas effect are continuously sought. With the growing concerns of greenhouse emissions, biomass is set to become an important contributor to the world energy need. Today biomass is seen as the most promising energy source to mitigate greenhouse gas emissions (Khan et al. 2009).

Biomass is very important for implementing the Kyoto agreement to reduce carbon dioxide emissions by replacing fossil fuels. The rapidly increasing energy requirements run parallel to technological development in the world, and research and development activities are forced to study new and biorenewable energy.

In the most biomass-intensive scenario, modernized biomass energy is projected by 2050 to contribute about one half of the total energy demand in developing countries. Direct combustion of biomass has been carried out worldwide since ancient times for cooking and heating. However, since biomass usually contains a high moisture content, and has low density, there are some difficulties of transportation, storage and usage of biomass without any pretreatment. Direct combustion is the old way of using biomass. Biomass thermochemical conversion technologies such as pyrolysis and gasification are certainly not the most important options at present; combustion is responsible for over 97% of the world's bioenergy production.

The first biomass sources used on Earth were wood and dry grass and for a long time these were used for cooking and warming. Products with fuel characteristics that are obtained biotechnologically from plant sources are defined as biomass energy sources. Generally biomass is an easily obtained energy source and therefore it is especially important for countries with forest and agriculture-based economics and with limited sources of energy (Karaosmanoglu and Aksoy 1988).

In the last decade, there has been rapid progress in the biofuel marketing trend: increasing production capacity, increasing international material flows, increased competition with conventional agriculture, increased competition with forest industries, increasing international trade flows, and strong international debate about the sustainability of biofuels production. Biomass is the most important bioenergy option at present and is expected to maintain that position during the first half of this century (Hamelinck and Faaij 2006; IPCC 2007).

There are three ways to use biomass: (1) it can be burned to produce heat and electricity, (2) changed to gas-like fuels such as methane, hydrogen and carbon monoxide, or (3) changed to a liquid fuel. Liquid fuels, also called biofuels, include mainly two forms of alcohol: ethanol and methanol. Because biomass can be changed directly into a liquid fuel, it could someday supply much of our transportation fuel needs for cars, trucks, buses, airplanes, and trains. This is very important because nearly one-third of our nation's energy is now used for transportation.

The term *biomass* (Greek *bio* meaning *life* + *maza* meaning *mass*) refers to non-fossilized and biodegradable organic material originating from plants, animals, and microorganisms. The biomass includes products, byproducts, residues and waste from agriculture, forestry and related industries as well as the non-fossilized and biodegradable organic fractions of industrial and municipal solid wastes. Biomass also includes gases and liquids recovered from the decomposition of non-fossilized and biodegradable organic material.

According to another definition, the term biomass refers to wood, short-rotation woody crops, agricultural wastes, short-rotation herbaceous species, wood wastes, bagasse, industrial residues, waste paper, municipal solid waste, sawdust, biosolids, grass, waste from food processing, aquatic plants and algae animal wastes, and a host of other materials.

Biomass is organic material that has stored sunlight in the form of chemical energy. Biomass is commonly recognized as an important renewable energy, which is considered to be such a resource that during the growth of plants and trees solar energy is stored as chemical energy via photosynthesis, which can be released via direct or indirect combustion.

Although some of the steps in photosynthesis are still not completely understood, the overall photosynthetic reaction has been known since the 1800s. Jan van Helmont began research into the process in the mid-1600s when he carefully measured the mass of the soil used by a plant and the mass of the plant as it grew. After noticing that the soil mass changed very little, he hypothesized that the mass of the growing plant must come from the water, the only substance he added to the potted plant. Photosynthesis is the most important biochemical process in which plants, algae, and some bacteria harness the energy of sunlight to produce food. Organisms that produce energy through photosynthesis are called photoautotrophs. Photosynthesis is a process in which green plants utilize the energy of sunlight to manufacture carbohydrates from carbon dioxide and water in the presence of chlorophyll (Viswanathan 2006).

The initial process in photosynthesis is the decomposition of water into oxygen and hydrogen and oxygen will be released. The simplified photosynthesis pathways are given in Eqs. 2.1–2.3. The hydrogen and the carbon and oxygen of carbon dioxide are then converted into formaldehyde (Eq. 2.1), and then a series of increasingly complex compounds result finally in a stable organic compound, hexose (typically glucose) (Eq. 2.2). This phase of photosynthesis utilizes stored energy and therefore

Table 2.1 Structural composition of wood (wt.% of dry and ash-free sample)

Wood species	Cellulose	Hemicelluloses	Lignin	Extractives
Hardwood	43–48	27–35	16–24	2–8
Softwood	40–44	24–29	26–33	1–5

can proceed in the dark. Hexose is polymerized into hexosan (typically glucosan) or carbohydrates (Eq. 2.3).

$$CO_2 + H_2O + \text{Solar energy} \xrightarrow{T>285\,\text{K, Chlorophyll}} CH_2O + O_2 \quad (2.1)$$

Glucose formation:

$$\underset{\text{Formaldehyde}}{6CH_2O} \rightarrow \underset{\text{Glucose}}{C_6H_{12}O_6} \quad (2.2)$$

Glucosan formation via polymerization:

$$\underset{\text{Glucose}}{nC_6H_{12}O_6} \rightarrow \underset{\text{Glucosan}}{(C_6H_{12}O_6)n} \quad (2.3)$$

The basic structure of all wood and woody biomass consists of cellulose, hemicelluloses, lignin and extractives. Their relative composition is shown in Table 2.1. Softwoods and hardwoods differ greatly in wood structure and composition. Hardwoods contain a greater fraction of vessels and parenchyma cells. Hardwoods have a higher proportion of cellulose, hemicelluloses and extractives than softwoods, but softwoods have a higher proportion of lignin. Hardwoods are denser than softwoods.

2.1.1 Biomass Feedstocks

Biomass feedstocks are marked by their tremendous diversity, which makes them rather difficult to characterize as a whole. Feedstocks that can be utilized with conversion processes are primarily the organic materials now being landfilled. These include forest products wastes, agricultural residues, organic fractions of municipal solid wastes, paper, cardboard, plastic, food waste, green waste, and other waste. Non-biodegradable organic feedstocks, such as most plastics, are not convertible by biochemical processes. Bio-based materials require pretreatment by chemical, physical, or biological means to open up the structure of biomass.

The major categories of biomass feedstock are as follows:

1. Forest products
   - Wood
   - Logging residues

  - Trees, shrubs and wood residues
  - Sawdust, barks, roots, etc.
2. Biorenewable wastes
   - Agricultural wastes
   - Crop residues
   - Mill wood wastes
   - Urban wood wastes
   - Urban organic wastes
   - Municipal solid wastes
   - Domestic solid wastes
3. Energy crops
   - Short-rotation woody crops
   - Herbaceous woody crops
   - Grasses
   - Starch crops
   - Sugar crops
   - Forage crops
   - Oilseed crops
4. Aquatic plants
   - Algae
   - Water weed
   - Water hyacinth
   - Reed and rushes
5. Food crops
   - Grains
   - Oil crops
6. Sugar crops
   - Sugar cane
   - Sugar beets
   - Molasses
   - Sorghum
7. Landfill
8. Industrial organic wastes
9. Algae and mosses
10. Kelps and lichens

Biomass is the world's fourth largest energy source worldwide, following coal, oil and natural gas. Biomass appears to be an attractive feedstock for three main reasons. First, it is a renewable resource that could be sustainably developed in the future. Second, it appears to have formidably positive environmental properties, reduced GHG emissions, possibly reduced NO_x and SO_x depending on the fossil fuels displaced. However, it also has negative impacts, such as polycyclic aromatic hydrocarbons including dioxins, furans, volatile organic compounds, and heavy metals especially when combusted in traditional stoves. Third, it appears to have significant economic potential provided that fossil fuel prices increase in the future.

Biomass for energy, especially biofuels, has positive attributes that contribute to a healthy environment and economy. Biomass utilization can reduce forest management costs, help mitigate climate change, reduce risks to life and property, and help provide a secure, competitive energy source. Biomass is a sustainable feedstock for chemicals and energy products that could potentially enhance the energy independence of the world, which lack energy resources. As an energy source that is highly productive, renewable, carbon neutral, and easy to store and transport, biomass has drawn worldwide attention recently.

Biomass offers important advantages as a combustion feedstock due to the high volatility of the fuel and the high reactivity of both the fuel and the resulting char. However, it should be noticed that in comparison with solid fossil fuels, biomass contains much less carbon and more oxygen and has a low heating value.

The waste products of a home include paper, containers, tin cans, aluminum cans, and food scraps, as well as sewage. The waste products of industry and commerce include paper, wood, and metal scraps, as well as agricultural waste products. Biodegradable wastes, such as paper fines and industrial biosludge, into mixed alcohol fuels (e.g., iso-propanol, iso-butanol, iso-pentanol). The wastes are first treated with lime to enhance reactivity; then they are converted to volatile fatty acids (VFAs) such as acetic acid, propionic acid, and butyric acid, using a mixed culture of microorganisms derived from cattle rumen or anaerobic waste treatment facilities. Pulp and paper wastes may also be treated to produce methane. The contents of domestic solid waste are given in Table 2.2 (Demirbas 2004).

Typical solid wastes include wood material, pulp and paper industry residues, agricultural residues, organic municipal material, sewage, manure, and food processing byproducts. Biomass is considered one of the main renewable energy resources of the future due to its large potential, economic viability and various social and environmental benefits. It was estimated that by 2050 biomass could provide nearly 38% of the world's direct fuel use and 17% of the world's electricity. If biomass is produced more efficiently and used with modern conversion

Table 2.2 Contents of domestic solid waste (wt.% of total)

Component	Lower limit	Upper limit
Paper waste	33.2	50.7
Food waste	18.3	21.2
Plastic matter	7.8	11.2
Metal	7.3	10.5
Glass	8.6	10.2
Textile	2.0	2.8
Wood	1.8	2.9
Leather and rubber	0.6	1.0
Miscellaneous	1.2	1.8

technologies, it can supply a considerable range and diversity of fuels at small and large scales. Municipal solid waste (MSW) is defined as waste durable goods, non-durable goods, containers and packaging, food scraps, yard trimmings, and miscellaneous inorganic wastes from residential, commercial, and industrial sources (Demirbas 2004). Figure 2.1 shows the use of world wood products (lumber, plywood, paperboard) between the years of 1970 and 2005.

Generally speaking, in producing fuels from biomass, one overall objective is to remove oxygen; biomass feedstocks often contain 40–60 wt.% oxygen and conventional fuels and oils typically have only trace amounts, under 1%. Oxygen heteroatom removal occurs most readily by dehydration, which removes oxygen in the form of water, and by decarboxylation, which removes oxygen in the form of carbon dioxide. Thermodynamically, since both water and carbon dioxide are fully oxidized and have no residual heating value, they can make ideal compounds in which to remove oxygen without losing heating value to the oxygen-containing chemicals removed (Peterson et al. 2008). Table 2.3 shows the elemental analyses of typical fuel samples given in the literature (Demirbas 2004).

The importance of biomass in different world regions is given in Table 2.4. As shown in this table, the importance of biomass varies significantly across regions. In Europe, North America and the Middle East, the share of biomass averages to 2–3% of total final energy consumption, whereas in Africa, Asia and Latin America, which together account for three-quarters of the world's population, biomass provides a substantial share of the energy needs: a third on average, but as much as 80–90% in some of the poorest countries of Africa and Asia (e. g., Angola, Ethiopia, Mozambique, Tanzania, Democratic Republic of Congo, Nepal, and Myanmar). Indeed, for large portions of the rural populations of developing countries, and for the poorest sections of urban populations, biomass is often the only available and affordable source of energy for basic needs such as cooking and heating.

Agricultural residues, grasses, algae, kelps, lichens and mosses are also important biomass feedstocks in the world. Algae can grow practically in every place

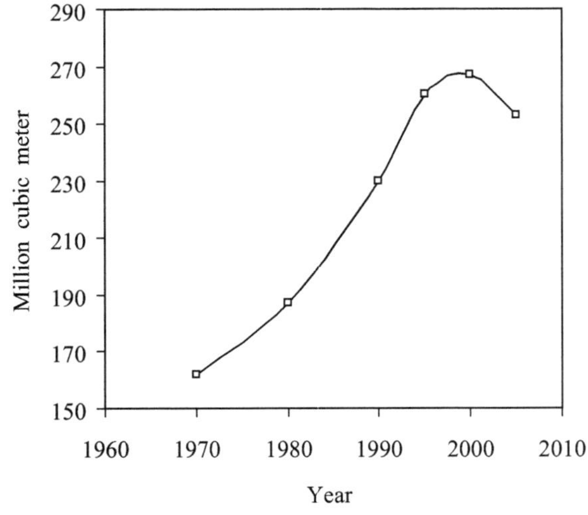

Fig. 2.1 Use of world wood products (1970–2005)

Table 2.3 Elemental analyses of typical fuel samples given in the literature (wt.% of dry fuel with ash)

Fuel sample	C	H	N	S	O (by difference)
Hazelnut shell	52.8	5.6	1.4	0.04	42.6
Sawdust	46.9	5.2	0.1	0.04	37.8
Corn stover	42.5	5.0	0.8	0.20	42.6
Red oak wood	50.0	6.0	0.3	–	42.4
Wheat straw	41.8	5.5	0.7	–	35.5
Olive husk	49.9	6.2	1.6	0.05	42.0
Poplar wood	48.4	5.9	0.4	0.01	39.6
Beech wood	49.5	6.2	0.4	–	41.2
Spruce wood	51.9	6.1	0.3	–	40.9
Corncob	49.0	5.4	0.5	0.20	44.5
Tea waste	48.0	5.5	0.5	0.06	44.0
Walnut shell	53.5	6.6	1.5	0.10	45.4
Almond shell	47.8	6.0	1.1	0.06	41.5
Sunflower shell	47.4	5.8	1.4	0.05	41.3
Rice husk	47.8	5.1	0.1	–	38.9
Cotton gin	42.8	5.4	1.4	0.50	35.0
Sugarcane bagasse	44.8	5.4	0.4	0.01	39.6
Peach pit	53.0	5.9	0.3	0.05	39.1
Alfalfa stalk	45.4	5.8	2.1	0.09	36.5
Switchgrass	46.7	5.9	0.8	0.19	37.4

where there is enough sunshine. Some algae can even grow in saline water. The most significant difference of algal oil is in the yield and hence its biodiesel yield. According to some estimates, the yield (per acre) of oil from algae is over 200 times the yield from the best-performing plant/vegetable oils (Sheehan et al. 1998). Microalgae are the fastest growing photosynthesizing organisms. They can complete an entire growing cycle every few days. Approximately 46 tons of oil per hectare per year can be produced from diatom algae. Different algae species produce different amounts of oil. Some algae produce up to 50% oil by weight.

Wood fuel is classified into three main commodities: fuelwood, charcoal, and black liquor. Black liquor is the spent pulping chemicals and the lignin component of wood after chemical pulping. It is fired in a chemical recovery boiler and process steam and electricity are also produced. Fuelwood and charcoal are the traditional wood forest products, and even today almost half of all of the forest harvest is for energy, with the remainder for industrial use.

Table 2.5 shows wood fuels consumption in 2005, given in petajoules (1 PJ = 10^9 MJ).

Table 2.4 The importance of biomass in different world regions

Region	Share of biomass in final energy consumption (%)
Africa (average)	62.0
Burundi	93.8
Ethiopia	85.6
Kenya	69.6
Somalia	86.5
Sudan	83.7
Uganda	94.6
South Asia (average)	56.3
East Asia (average)	25.1
China	23.5
Latin America (average)	18.2
Europe (average)	3.5
North America (average)	2.7
Middle East (average)	0.3

Table 2.5 Wood fuel consumption in 2005 (1 PJ = 10^9 MJ)

	Fuelwood	Black liquor	Charcoal
Asia	7795	463	135
Africa	5633	33	688
Latin America	2378	288	485
Europe	1173	644	14
North America	852	1284	40
Oceania	90	22	1
Total	17 921	2734	1361

2.1.2 Modern Biomass, Bioenergy, and Green Energy

The term *modern biomass* is generally used to describe the traditional biomass used through efficient and clean combustion technologies and a sustained supply of biomass resources, environmentally sound and competitive fuels, heat and electricity using modern conversion technologies. Modern biomass produced in a sustainable way excludes traditional uses of biomass as fuelwood and includes electricity generation and heat production, as well as transportation fuels, from agricultural and forest residues and solid waste. On the other hand, *traditional biomass* is produced in an unsustainable way and it is used as a non-commercial source – usually with very low efficiencies for cooking in many countries (Goldemberg and Coelho 2004).

Traditional biomass markets have been inefficient but technological developments have reduced energy, emission, and material flows through the system thus improving the efficiency of biomass energy systems. The energy market demands cost effectiveness, high efficiency, and reduced risk to future emission limits. Modernization of biomass conversion implies the choice of technologies: (a) offers the potential for high yields, (b) economic fuel availability, (c) low adverse environmental impacts, and (d) suitability to modern energy systems. A number of systems that meet the mentioned criteria for modernized biomass conversion can be identified (Larson 1993).

Bioenergy is energy derived from organic sources such as all lignocellulosic materials, trees, agricultural crops, municipal solid wastes, food processing and agricultural wastes and manure. Generally, bioenergy is an inclusive term for all forms of biomass and biofuels.

Bioenergy is one of the forms of renewable energy. Biomass has been used for bioenergy for thousands of years, ever since people started burning wood to cook food and today, wood is still our largest biomass resource for bioenergy. The use of bioenergy has the potential to greatly reduce our greenhouse gas emissions. Replacing fossil fuels with energy from biomass has several distinct environmental implications. If biomass is harvested at a rate, which is sustainable, using it for energy purposes does not result in any net increase in atmospheric carbon dioxide, a greenhouse gas.

Green energy is an alternate term for renewable energy and refers to the energy generated from sources that are considered environmentally friendly (e.g., hydro, solar, biomass (landfill) or wind). Green power is sometimes used in reference to electricity generated from "green" sources. Green energy production is the principal contributor to the economic development of a developing country. Its economy development is based on agricultural production and most people live in the rural areas. Implementation of integrated community development programs is therefore very necessary.

Green power refers to electricity supplied from more readily renewable energy sources than traditional electrical power sources. Green power products have become widespread in many electricity markets worldwide, which can be derived from renewable energy sources. The environmental advantages of the production and use of green electricity seem to be clear. Market research indicates that there is a large potential market for green energy in Europe in general. Green power marketing has emerged in more than a dozen countries around the world.

Many green electricity products are based on renewable energy sources like wind, biomass, hydro, biogas, solar, and geothermal (Murphya and Niitsuma 1999). There has been interest in electricity from renewable sources named green electricity or green pool, as a special market (Elliott 1999). The term green energy is also used for green energy produced from co-generation, energy from municipal waste, natural gas, and even conventional energy sources. Using green energy sources like hydro, biomass, geothermal, and wind energy in electricity production reduces CO_2 emissions (Fridleifsson 2003). Emissions such as SO_2, CO_2 and NO_x are reduced considerably and the production and use of green electricity contributes to diminishing the green house effect (Arkesteijn and Oerlemans 2005).

Table 2.6 Electricity production from biomass (TWh)

	1995	2002	2003	2004	2005	2006
Solid biomass	85.3	110.0	118.2	131.4	134.9	140.6
Biogas	6.0	16.9	18.3	20.7	24.8	27.3
Liquid biomass			0.8	0.6	0.9	1.3
Municipal solid waste	13.4	21.3	25.0	24.0	22.8	21.8
Total	104.8	148.2	162.2	176.6	183.4	190.9

In general, a sustainable energy system includes energy efficiency, energy reliability, energy flexibility, energy development and continuity, combined heat and power (CHP) or co-generation, fuel poverty, and environmental impacts. The environmental impacts of energy use are not new. For centuries, wood burning has contributed to the deforestation of many areas. On the other hand, the typical characteristics of a sustainable energy system can be derived from political definitions. A sustainable energy system can be defined also by comparing the performance of different energy systems in terms of sustainability indicators (Alanne and Sari 2006). By definition, sustainable energy systems must support both human and ecosystem health over the long term, therefore goals for tolerable emissions should look well into the future. They should also take into account the public's tendency to demand more (UNDP 2000).

Various scenarios have been put forward on estimates of biofuel from biomass sources in the future energy system. The availability of the resources is an important factor in co-generative use of biofuel in the electricity, heat, or liquid fuel market. The largest secondary transformation of biomass after charcoal production is in the electricity sector. For many years biomass-processing industries such as sugar, wood products, and chemical pulping have installed combined heat and power.

In 2006 estimated total electricity generation was about 190 TWh from an installed capacity in excess of 43 GW. The recent negative trend for MSW is a consequence of increased material recycling, together with reductions in the amount of biomass-derived materials entering the waste stream. Table 2.6 shows the electricity production from biomass. Table 2.7 shows the leading biopower-producing countries in 2005 (WEC 2007). The eight leading countries with biomass-based electricity production are all members of the OECD, except for Brazil.

2.2 The Chemistry of Biomass

The components of biomass include cellulose, hemicelluloses, lignin, extractives, lipids, proteins, simple sugars, starches, water, hydrocarbons, ash, and other compounds. Two larger carbohydrate categories that have significant value are cellulose and hemicelluloses (holocellulose). The lignin fraction consists of non-sugar type

Table 2.7 Leading biopower producing countries in 2005

Country	Production (TWh)	Percentage of global total
United States	56.3	30.7
Germany	13.4	7.3
Brazil	13.4	7.3
Japan	9.4	5.1
Finland	8.9	4.9
UK	8.5	4.7
Canada	8.5	4.6
Spain	7.8	4.3
Rest of World	57.1	31.1

macromolecules. Three structural components are cellulose, hemicelluloses, and lignin, which have rough formulae of $CH_{1.67}O_{0.83}$, $CH_{1.64}O_{0.78}$, and $C_{10}H_{11}O_{3.5}$, respectively.

Cellulose is a linear polymer composed of repeating anhydroglucose units. Cellulose is a remarkable pure organic polymer, consisting solely of units of anhydroglucose held together in a giant straight-chain molecule (Demirbas 2000). These anhydroglucose units are bound together by β-(1,4)-glycosidic linkages. Due to this linkage, cellobiose is established as the repeat unit for cellulose chains. Cellulose must be hydrolyzed to glucose before fermentation to ethanol.

The second major chemical species in wood are the hemicelluloses. They are amorphous polysaccharides, such as xylans, galactoglucomannans, arabinogalactans, glucans and galactans. The hemicelluloses, unlike cellulose, not only contain glucose units, but they are also composed of a number of different pentose and hexose monosaccharides. Hemicelluloses tend to be much shorter in length than cellulose, and the molecular structure is slightly branched.

Unlike cellulose, hemicelluloses consist of different monosaccharide units. In addition, the polymer chains of hemicelluloses have short branches and are amorphous. Because of the amorphous morphology, hemicelluloses are partially soluble or swellable in water. Hemicelluloses (arabinoglycuronoxylan and galactoglucomannans) are related to plant gums in composition, and occur in much shorter molecule chains than cellulose. Hemicelluloses are derived mainly from chains of pentose sugars, and act as the cement material holding together the cellulose micelles and fiber (Theander 1985).

Lignin (sometimes lignen) is a chemical compound that is most commonly derived from wood and is an integral part of the cell walls of plants, especially in tracheids, xylem fibers, and sclereids. It is one of most abundant organic compounds on Earth after cellulose and chitin. The empirical chemical formula of lignin is $C_{20}H_{19}O_{14}N_2$. Lignin is a complex, high molecular weight polymer built of hydroxyphenylpropane units. Lignin is a completely different polymeric material, being highly cross-linked and having phenolic-like structures as the monomeric base. It is the lignin that holds the wood cells together and provides the extraordinary composite strength characteristics of a piece of wood.

Lignin is a large, cross-linked macromolecule with molecular mass in excess of 10 000 amu. It is relatively hydrophobic and aromatic in nature. The molecular weight in nature is difficult to measure, since it is fragmented during preparation. The molecule consists of various types of substructures, which appear to repeat in a random manner. Lignin plays a crucial part in conducting water in plant stems. The polysaccharide components of plant cell walls are highly hydrophilic and thus permeable to water, whereas lignin is more hydrophobic. The cross-linking of polysaccharides by lignin is an obstacle for water absorption to the cell wall. Thus, lignin makes it possible for plants to form vascular tissue, which conducts water efficiently. Lignin is present in all vascular plants, but not in bryophytes, supporting the idea that the main function of lignin is related to water transport.

Lignins are polymers of aromatic compounds. Their functions are to provide structural strength, provide sealing of water-conducting systems that link roots with leaves, and protect plants against degradation (Glasser and Sarkanen 1989). Lignin is a macromolecule, which consists of alkyl phenols and has a complex three-dimensional structure. Lignin is covalently linked with xylans in the case of hardwoods and with galactoglucomannans in softwoods (Sarkanen and Ludwig 1971).

Pyrolysis of lignin, e. g., during combustion, yields a range of products, of which the most characteristic ones are methoxy phenols. Of those, the most important are guaiacol and syringol and their derivatives; their presence can be used to trace a smoke source to a wood fire. Lignin biosynthesis begins with the synthesis of monolignols. The starting material is the amino acid phenylalanine. The first reactions in the biosynthesis are shared with the phenylpropanoid pathway, and monolignols are considered to be a part of this group of compounds. There are three main types of monolignols: coniferyl alcohol, sinapyl alcohol and paracoumaryl alcohol. Different plants use different monolignols. For example, gymnosperms as Norway spruce have a lignin that consists almost entirely of coniferyl alcohol. Dicotyledonic lignin is a mixture of coniferyl alcohol and sinapyl alcohol (normally more of the latter), and monocotyledonic lignin is a mixture of all three monolignols. Some monocotyledons have mostly coniferyl alcohol (as many grasses), while others have mainly sinapyl alcohols, as some palms. Monolignols are synthesized in the cytosol as glucosides. The glucose is added to the monolignol to make them water-soluble and to reduce their toxicity. The glucose is then removed and the monolignols are polymerized into lignin. Four main monolignols in lignin structure are given in Fig. 2.2.

Wood and woody biomass also contain lesser amounts of tannins, simple sugars, starches, pectins and organic soluble extractives. Extractives include terpenes, tall oil and the fatty acids, esters, and triglycerides, which contribute to paper mill pitch problems (Demirbas 1991).

Characterization of biomass fuels has been reviewed by Bushnell et al. (1989). In general combustion models of biomass can be classified as macroscopic or microscopic. The characterization of the biomass fuels is generally divided into three broad categories: (1) composition and structure, (2) reactivity measures, and (3) ash chemistry.

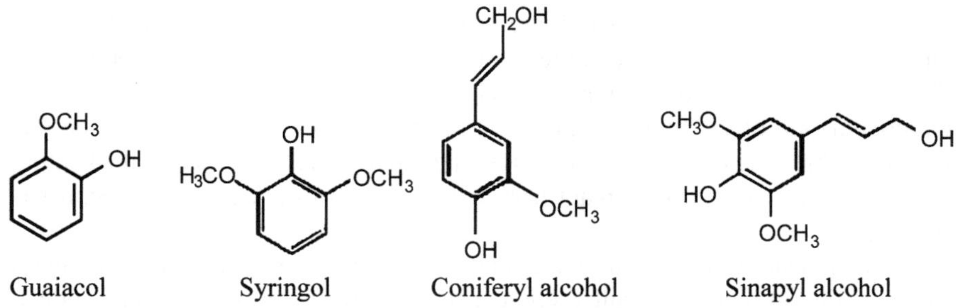

Fig. 2.2 Four main monolignols in lignin structure

Properties for microscopic analysis include thermal, chemical kinetic and mineral data (Ragland et al. 1991). Fuel characteristics such as ultimate analysis, heating value, moisture content, particle size, bulk density, and ash fusion temperature of biomass have been reviewed (Bushnell et al. 1989). Fuel properties for the combustion analysis of biomass can be conveniently grouped into physical, chemical, thermal, and mineral properties.

Physical property values vary greatly and properties such as density, porosity, and internal surface area are related to biomass species whereas bulk density, particle size, and shape distribution are related to fuel preparation methods. Important chemical properties for combustion are the ultimate analysis, proximate analysis, analysis of pyrolysis products, higher heating value, heat of pyrolysis, heating value of the volatiles, and heating value of the char.

Thermal property values such as specific heat, thermal conductivity, and emissivity vary with moisture content, temperature, and degree of thermal degradation by one order of magnitude. Thermal degradation products of biomass consist of moisture, volatiles, char and ash. Volatiles are further subdivided into gases such as light hydrocarbons, carbon monoxide, carbon dioxide, hydrogen and moisture, and tars. The yields depend on the temperature and heating rate of pyrolysis. Some properties vary with species, location within the biomass, and growth conditions. Other properties depend on the combustion environment. Where the properties are highly variable, the likely range of the property is given (Ragland et al. 1991).

2.2.1 Biomass Fuel Analyses

Main biomass fuel analyses are: (a) particle size and specific gravity, (b) ash content, (c) moisture content, (d) extractive content, (e) element (C, H, O and N) content, and (f) structural constituent (cellulose, hemicelluloses and lignin) content.

Particle size of biomass should be as much as 0.6 cm, sometimes more, in a profitable combustion process. Biomass is much less dense and has significantly higher aspect ratios than coal. It is also much more difficult to reduce to small sizes.

Ash or inorganic material in plants depends on the type of the plant and the soil contamination in which the plant grows. On average wood contains about 0.5% ash.

Ash contents of hard and softwoods are about 0.5 and 0.4%, respectively. Insoluble compounds act as a heat sink in the same way as moisture, lowering combustion efficiency, but soluble ionic compounds can have a catalytic effect on the pyrolysis and combustion of the fuel. The presence of inorganic compounds favors the formation of char. Ash content is an important parameter directly affecting the heating value. High ash content of a plant part makes it less desirable as fuel (Demirbas 1998).

The composition of mineral matter can vary between and within each biomass sample. Mineral matter in fruit shells consists mostly of salts of calcium, potassium, silica, and magnesium, but salts of many other elements are also present in lesser amounts (Demirbas 2002).

Moisture in biomass generally decreases its heating value. Moisture in biomass is stored in spaces within the dead cells and within the cell walls. When the fuel is dried the stored moisture equilibrates with the ambient relative humidity. Equilibrium is usually about 20% in air-dried fuel. Moisture percentage of the wood species varies from 41.27 to 70.20%. Heating value of a wood fuel decreases with increase in moisture content of the wood. Moisture content varies from one tree part to another. It is often the lowest in the stem and increases toward the roots and the crown. The presence of water in biomass influences its behavior during pyrolysis and affects the physical properties and quality of the pyrolysis liquid. The results obtained show that for higher initial moisture contents the maximum liquid yield on a dry feed basis occurs at lower pyrolysis temperatures between 691 and 702 K (Demirbas 2004).

Again, the heat content, which is a very important factor affecting the utilization of any material as a fuel, is affected by the proportion of combustible organic components (known as extractives) present in it. The higher heating values (HHVs) of the extractive-free plant parts were found to be lower than those of the unextracted parts, which indicate a likely positive contribution of extractives towards the increase of HHV. Extractive content is an important parameter directly affecting the heating value. High extractive content of a plant part makes it desirable as fuel. Extractives raise the HHVs of the wood fuels.

Both the chemical and the physical composition of the fuel are important factors in the characteristics of combustion. Biomass can be analyzed by breaking it down into structural components (called proximate analysis) or into chemical elements (known as ultimate analysis). The heat content is related to the oxidation state of the natural fuels in which carbon atoms generally dominate and overshadow small variations of hydrogen content. On the basis of literature values for different species of wood, Tillman (1978) also found a linear relationship between HHV and carbon content.

Biomass fuels are composed of biopolymers that consist of various types of cells and the cell walls are built of cellulose, hemicelluloses and lignin. HHVs of biomass fuels increase as lignin contents increase.

The higher heating values (HHVs) or gross heat of combustion includes the latent heat of the water vapor products of combustion because the water vapor was allowed to condense to liquid water. The HHV (in units ofMJ/kg) of the biomass fuels as a function of fixed carbon (FC, wt.%) was calculated from Eq. 2.1 (Demirbas 1997):

$$HHV = 0.196\,(FC) + 14.119 \quad (2.4)$$

In earlier work (Demirbas et al. 1997), formulae were also developed for estimating the HHVs of fuels from different lignocellulosic materials, vegetable oils, and diesel fuels using their chemical analysis data. For biomass fuels such as coal, the HHV is calculated using a modified Dulong's formula as a function of the carbon, hydrogen, oxygen, and nitrogen contents from Eq. 2.2:

$$HHV = 0.335\,(CC) + 1.423\,(HC) - 0.154\,(OC) - 0.145\,(NC) \quad (2.5)$$

where CC is carbon content (wt.%), HC hydrogen content (wt.%), OC oxygen content (wt.%), and NC nitrogen content (wt.%).

The HHVs of extractive-free samples reflect the HHV of lignin relative to cellulose and hemicelluloses. Cellulose and hemicelluloses (holocellulose) have a HHV 18.60 MJ/kg, whereas lignin has a HHV from 23.26 to 26.58 MJ/kg. As discussed by Baker (1982), HHVs reported for a given species reflect only the samples tested and not the entire population of the species. The HHV of a lignocellulosic fuel is a function of its lignin content. In general, the HHVs of lignocellulosic fuels increase with increase of their lignin contents and the HHV is highly correlated with lignin content. For the model including the lignin content, the regression equation is:

$$HHV = 0.0889\,(LC) + 16.8218 \quad (2.6)$$

where LC was the lignin content (wt.% dry and ash-free and extractive-free basis).

Biomass combustion is a series of chemical reactions by which carbon is oxidized to carbon dioxide, and hydrogen is oxidized to water. Oxygen deficiency leads to incomplete combustion and the formation of many products of incomplete combustion. Excess air cools the system. The air requirements depend on the chemical and physical characteristics of the fuel. The combustion of the biomass relates to the fuel burn rate, the combustion products, the required excess air for complete combustion, and the fire temperatures.

2.3 Biomass Conversion Processes

Biomass can be converted into useful biofuels and biochemicals via biomass upgrading and biorefinery technologies. Biomass conversion processes include fractionation, densification (briquetting, pelleting), liquefaction, supercritical fluid liquefaction, destructive carbonization, pyrolysis, gasification, hydrothermal liquefaction and hydrothermal upgrading, Fischer–Tropsch synthesis, anaerobic digestion, hydrolysis, and fermentation. Figure 2.3 shows the main biomass conversion processes.

Direct combustion is the oldest way of using biomass. Biomass thermochemical conversion technologies such as pyrolysis and gasification are certainly not the

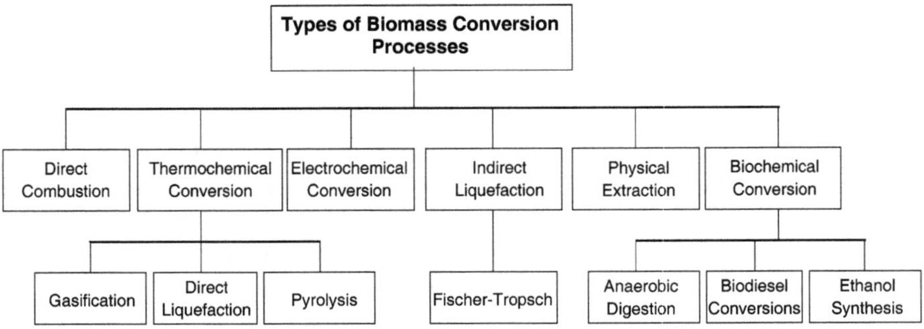

Fig. 2.3 Main biomass conversion processes

most important options at present; combustion is responsible for over 97% of the world's bioenergy production (Demirbas 2004). Direct combustion and co-firing with coal for electricity production from biomass has been found to be a promising method for the near future. The supply is dominated by traditional biomass used for cooking and heating, especially in rural areas of developing countries. The traditional biomass cooking and heating produces high levels of pollutants.

Biomass energy currently represents approximately 14% of total global energy consumption, a higher share than that of coal (12%) and comparable to those of gas (15%) and electricity (14%). Biomass is the main source of energy for many developing countries and most of it is non-commercial. Hence there is enormous difficulty in collecting reliable biomass energy data. Yet good data are essential for analyzing tendencies and consumption patterns, for modeling future trends and for designing coherent strategies.

The three ways to use biomass are: (1) burning it to produce heat and electricity, (2) changing it to gas-like fuels such as methane, hydrogen and carbon monoxide, and (3) changing it to a liquid fuel. Liquid fuels, also called biofuels, include ethanol and methanol. Ethanol, the most commonly used biofuel, is produced from sugarcane, corn, and other grains. A blend of gasoline and ethanol is already used in cities with high levels of air pollution.

There are several ways to make use of the energy contained in the biomass from direct burning to pyrolysis, gasification, and liquefaction. Pyrolysis is the thermochemical process that converts organic materials into usable fuels. Pyrolysis produces energy fuels with high fuel-to-feed ratios, making it the most efficient process for biomass conversion and the method most capable of competing with and eventually replacing non-renewable fossil fuel resources.

Main transportation fuels can be obtained from biomass using different processes, fuels such as sugar ethanol, cellulosic ethanol, grain ethanol, biodiesel, pyrolysis liquids, green diesel, green gasoline, buthanol, methanol, syngas liquids, biohydrogen, algae diesel, algae jet fuel, and hydrocarbons.

In the liquefaction process, biomass is converted to liquefied products through a complex sequence of physical structure and chemical changes. The feedstock of liquefaction is usually a wet matter. In the liquefaction process, biomass is decomposed into small molecules. These small molecules are unstable and reactive, and can repolymerize into oily compounds with a wide range of molecular weight distribution (Demirbas 2000).

Liquefaction can be accomplished directly or indirectly. Direct liquefaction involves rapid pyrolysis to produce liquid tars and oils and/or condensable organic vapors. Indirect liquefaction involves the use of catalysts to convert non-condensable, gaseous products of pyrolysis or gasification into liquid products. Alkali salts, such as sodium carbonate and potassium carbonate, can act as catalysts for the hydrolysis of cellulose and hemicellulose, into smaller fragments. The degradation of biomass into smaller products mainly proceeds by depolymerization and deoxygenation. In the liquefaction process, the amount of solid residue increases in proportion to the lignin content. Lignin is a macromolecule consisting of alkyl phenols, and has a complex three-dimensional structure. It is generally accepted that free phenoxyl radicals are formed by thermal decomposition of lignin above 500 K and that the radicals have a random tendency to form a solid residue through condensation or repolymerization (Demirbas 2000).

The changes during liquefaction process involve all kinds of processes such as solvolysis, depolymerization, decarboxylation, hydrogenolysis, and hydrogenation. Solvolysis results in micellar-like substructures of the biomass. The depolymerization of biomass leads to smaller molecules. It also leads to new molecular rearrangements through dehydration and decarboxylation. When hydrogen is present, hydrogenolysis and hydrogenation of functional groups, such as hydroxyl groups, carboxyl groups, and keto groups also occur.

Pyrolysis is the basic thermochemical process for converting biomass to a more useful fuel. Biomass is heated in the absence of oxygen, or partially combusted in a limited oxygen supply, to produce a hydrocarbon-rich gas mixture, an oil-like liquid and a carbon-rich solid residue. Rapid heating and rapid quenching produce the intermediate pyrolysis liquid products, which condense before further reactions break down higher molecular weight species into gaseous products. High reaction rates minimize char formation. Under some conditions, no char is formed. At higher fast pyrolysis temperatures, the major product is gas.

Pyrolysis is the simplest and almost certainly the oldest method of processing one fuel in order to produce a better one. Pyrolysis can also be carried out in the presence of a small quantity of oxygen (gasification), water (steam gasification) or hydrogen (hydrogenation). One of the most useful products is methane, which is a suitable fuel for electricity generation using high-efficiency gas turbines.

Cellulose and hemicelluloses form mainly volatile products on heating due to the thermal cleavage of the sugar units. The lignin forms mainly char since it is not readily cleaved to lower molecular weight fragments. The progressive increase in the pyrolysis temperature of the wood led to the release of the volatiles thus forming a solid residue that is different chemically from the original starting material (Demirbas 2000). Cellulose and hemicelluloses initially break into compounds of lower molecular weight. This forms an activated cellulose, which decomposes by two competitive reactions: one forming volatiles (anhydrosugars), and the other char and gases. The thermal degradation of the *activated cellulose* and hemicelluloses to form volatiles and char can be divided into categories depending on the reaction temperature. Within a fire all these reactions take place concurrently and consecutively. Gaseous emissions are predominantly a product of pyrolitic cracking of the fuel. If flames are present, fire temperatures are high, and more oxygen is available from thermally induced convection.

The biomass pyrolysis is attractive because solid biomass and wastes can be readily converted into liquid products. These liquids, as crude bio-oil or slurry of charcoal of water or oil, have advantages in many fields such as transport, storage, combustion, retrofitting and flexibility in production and marketing.

Gasification is a form of pyrolysis, carried out in the presence of a small quantity of oxygen at high temperatures in order to optimize the gas production. The resulting gas, known as *producer gas*, is a mixture of carbon monoxide, hydrogen and methane, together with carbon dioxide and nitrogen. The gas is more versatile than the original solid biomass (usually wood or charcoal): it can be burnt to produce process heat and steam, or used in gas turbines to produce electricity.

Biomass gasification technologies are expected to be an important part of the effort to meet these goals of expanding the use of biomass. Gasification technologies provide the opportunity to convert renewable biomass feedstocks into clean fuel gases or synthesis gases. Biomass gasification is the latest generation of biomass energy conversion processes, and is being used to improve the efficiency, and to reduce the investment costs of biomass electricity generation through the use of gas turbine technology. High efficiencies (up to about 50%) are achievable using combined-cycle gas turbine systems, where waste gases from the gas turbine are recovered to produce steam for use in a steam turbine.

Various gasification technologies include gasifiers where the biomass is introduced at the top of the reactor and the gasifying medium is either directed co-currently (downdraft) or counter-currently up through the packed bed (updraft). Other gasifier designs incorporate circulating or bubbling fluidized beds. Tar yields can range from 0.1% (downdraft) to 20% (updraft) or greater in the product gases.

The process of synthetic fuels (synfuels) from biomass will lower the energy cost, improve the waste management, and reduce harmful emissions. This triple assault on plant operating challenges is a proprietary technology that gasifies biomass by reacting it with steam at high temperatures to form a clean burning synthetic gas (syngas: $CO + H_2$). The molecules in the biomass (primarily carbon, hydrogen and oxygen) and the molecules in the steam (hydrogen and oxygen) reorganize to form this syngas.

2.4 Chemicals from Biomass

Such features of larch wood, as an increased content of extractive compounds and its high density, create some technological problems for the pulping process. It seems that production of high-value-added chemicals is the most profitable from an economical point of view of larch wood valorization. High-value organic compounds such as arabinogalactan, quercitin dihydrate vanillin, microcrystalline cellulose and levulinic acid are obtained from larch wood. Both arabinogalactan and quercitin dihydrate can be extracted from larch wood with boiling water (Kuznetsov et al. 2005).

Chemicals can be obtained from thermal depolymerization and decomposition of biomass structural components; cellulose, hemicelluloses, and lignin form liquids and gas products as well as a solid residue of charcoal. The bio-oils from pyrolysis of biomass are composed of a range of cyclopentanone, methoxyphenol, acetic acid, methanol, acetone, furfural, phenol, formic acid, levoglucosan, guaiacol and their alkylated phenol derivatives. The structural components of the biomass samples mainly have an affect on the pyrolytic degradation products. A reaction mechanism is proposed which describes a possible reaction route for the formation of the characteristic compounds found in the oils. The temperature and heating rate are the most important parameters, which affect the composition of chemicals. The supercritical water conditioning and liquefaction partial reactions also occur during the pyrolysis. Acetic acid is formed in the thermal decomposition of all three main components of biomass. In the pyrolysis reactions of biomass water is formed by dehydration, acetic acid comes from the elimination of acetyl groups originally linked to the xylose unit, furfural is formed by dehydration of the xylose unit, formic acid proceeds from carboxylic groups of uronic acid, and methanol arises from methoxyl groups of uronic acid.

The pyrolysis process can produce mainly charcoal, condensable organic liquids, non-condensable gases, acetic acid, acetone, and methanol. Among the liquid products, methanol is one of the most valuable products. The liquid fraction of the pyrolysis products consists of two phases: an aqueous phase containing a wide variety of organo-oxygen compounds of low molecular weight and a non-aqueous phase containing insoluble organics of high molecular weight. This phase is called tar and is the product of greatest interest. The ratios of acetic acid, methanol, and acetone of the aqueous phase are higher than those of the non-aqueous phase.

Chemicals potentially derived from lignin conversions are syngas, methanol, dimethyl ether, ethanol, mixed alcohols, byproduct C_1 to C_4 gases, hydrocarbons, oxygenates, Fischer–Tropsch liquids, cyclohexane, styrenes, biphenyls, phenol, substituted phenols, catechols, cresols, resorcinols, eugenol, syringols, guaiacols, vanillin, vanilic acid, aromatic acids, aliphatic acids, syringaldehyde and aldehydes, quinones, cyclohexanol/al, cyclohexanal, beta keto adipate, benzene, toluene, xylene (BTX) and their derivates, higher alkylates, substituted lignins, drugs, mixed aromatic polyols, carbon fiber, fillers, etc.

As the temperature is increased there is an increase in the yield of gas. The influence of residence time on the yield of products is small, with a slight decrease in oil yield and increase in char yield. The main gases produced are carbon dioxide, carbon monoxide, methane and hydrogen and there is significant production of oil and char. Pyrolysis of biomass produces high conversion rates to a gas composed mainly of hydrogen and carbon dioxide within addition carbon monoxide and C_1–C_4 hydrocarbons. Similar to conventional gasification processes, oil and char reaction products are produced. The oils are composed of a range of oxygenated compounds, including, cyclopentanone, methoxybenzene, acetic acid, furfural, acetophenone, phenol, benzoic acid and their alkylated derivatives.

Biological processes are essentially microbic digestion and fermentation. High-moisture herbaceous plants (vegetables, sugar cane, sugar beet, corn, sorghum, and cotton), marine crops, and manure are most suitable for biological digestion. Intermediate heat gas is methane mixed with carbon monoxide and carbon dioxide. Methane (high heat gas) can be efficiently converted into methanol.

Cellulose is a remarkably pure organic polymer, consisting solely of units of anhydroglucose held together in a giant straight-chain molecule. Cellulose must be hydrolyzed to glucose before fermentation to ethanol. Conversion efficiencies of cellulose to glucose may be dependent on the extent of chemical and mechanical pretreatments to structurally and chemically alter the pulp and paper mill wastes. The method of pulping, the type of wood, and the use of recycled pulp and paper products also could influence the accessibility of cellulose to cellulase enzymes. Hemicelluloses (arabinoglycuronoxylan and galactoglucomannans) are related to plant gums in composition, and occur in much shorter molecule chains than cellulose. The hemicelluloses, which are present in deciduous woods chiefly as pentosans and in coniferous woods almost entirely as hexosanes, undergo thermal decomposition very readily. Hemicelluloses are derived mainly from chains of pentose sugars, and act as the cement material holding together the cellulose micelles and fiber. Cellulose is insoluble in most solvents and has a low accessibility to acid and enzymatic hydrolysis. Hemicelluloses are largely soluble in alkali and, as such, are more easily hydrolyzed.

Hydrolysis (saccharification) breaks down the hydrogen bonds in the hemicellulose and cellulose fractions into their sugar components: pentoses and hexoses. These sugars can then be fermented into bioethanol.

Cellulose (hexosan ($(C_6H_{10}O_5)_n$) hydrolysis produces glucose (a hexose, $C_6H_{12}O_6$). The hydrolysis of cellulose is catalyzed by mineral acids and enzymes. Hemicelluloses hydrolysis produces both hexose and pentose sugars: mannose, galactose, xylose and arabinose that are not all fermented with existing strains.

2.4.1 Char from Biomass

Agricultural and forestry byproducts may also offer an inexpensive and renewable additional source of activated carbons. These waste materials have little or no economic value and often present a disposal problem. Therefore, there is a need to valorize these low-cost byproducts. Therefore, their conversion into activated carbons would add economic value, help reduce the cost of waste disposal, and most importantly provide a potentially inexpensive alternative to the existing commercial activated carbons. Activated carbon is prepared from dried municipal sewage sludge and batch mode adsorption experiments (Reddy et al. 2006).

Active carbons are carbonaceous materials with highly developed internal surface area and porosity. It is widely used as an effective adsorbent in many applications such as air separation and purification, vehicle exhaust emission control, solvent recovery, and catalyst support because of its high specific pore surface area, adequate pore size distribution, and relatively high mechanical strength. The large surface area results in high capacity for adsorbing chemicals from gases and liquids (Zanzi 2001).

The starting materials used in commercial production of activated carbons are those with high carbon contents such as wood, lignite, peat, and coal of different ranks, or low-cost and abundantly available agricultural byproducts. Active carbons can be manufactured from virtually any carbonaceous precursor, but the most commonly used materials are wood, coal, and coconut shell. The development of activated carbons from agricultural carbonaceous wastes will be advantageous for environmental problems. In water contamination, wastewater contains many traces of organic compounds, which are a serious environmental problem. In the development of activated carbons, agricultural carbonaceous wastes will be used, as this will eliminate the problem of waste disposal while at the same time societies will derive great economic benefits from such commercialized products.

Activated carbons are used in the following applications:

1. They can be used as adsorbents for the removal or reduction of gaseous pollutants from the exhaust gases of industrial sources.
2. They can be used as adsorbents for the removal of volatile organic compounds, ozone from air, mercury and dioxin emissions from incinerator flue gas, and hydrogen sulfide emissions from sewage treatment facilities.
3. They can be used to remove chlorine and organic chemicals (such as phenols, polychlorinated biphenyls, trihalomethanes, pesticides and halogenated hydrocarbons), heavy metals and organic contaminants from water.
4. They can be used to extract some harmful elements of cigarette smoke by incorporation in filter tips of the cigarettes.

Lignin gives higher yields of charcoal and tar from wood, although lignin has a threefold higher methoxyl content than wood (Demirbas 2000). Phenolics are derived from lignin by cracking the phenyl-propane units of the macromolecule lattice. The formation of char from lignin under mild reaction conditions is a result of the breaking of the relatively weak bonds, like the alkyl-aryl ether bonds, and the consequent formation of more resistant condensed structures, as has already been noted (Domburg et al. 1974). One additional parameter that may also have an effect on the char formation is the moisture content of the kraft lignin used. It has been found that the presence of moisture increased the yield of char from the pyrolysis of wood waste at temperatures between 660 and 730 K (Demirbas 2006).

The destructive reaction of cellulose is started at temperatures lower than 425 K and is characterized by a decreasing polymerization degree. Thermal degradation of cellulose proceeds through two types of reaction: (1) a gradual degradation, decomposition and charring on heating at lower temperatures, and (2) a rapid volatilization accompanied by the formation of levoglucosan on pyrolysis at higher temperatures. The degradation of cellulose to a more stable anhydrocellulose, which gives higher biochar yield, is the dominant reaction at temperatures less than 575 K (Shafizadeh 1985). At temperatures greater than 575 K, cellulose depolymerizes, producing volatiles. If the heating rate is very high, the residence time of the biomass at temperatures less than 575 K is insignificant. Thus a high heating rate provides a shorter time for the dehydration reactions and the formation of less reactive anhydrocellulose, which gives a higher yield of char (Zanzi 2001). The result is that the rapid heating of the biomass favors the polymerization of cellulose and the formation of volatiles and suppresses the dehydration to anhydrocellulose and char formation. Hence the effect of heating rate is stronger in the pyrolysis of biomass than in coal.

The initial degradation reactions include depolymerization, hydrolysis, oxidation, dehydration, and decarboxylation (Shafizadeh 1982). The isothermal pyrolysis of cellulose in air and milder conditions, in the temperature range 623–643 K, was investigated (Fengel and Wegener 1983). Under these conditions, the pyrolysis reactions produced 62–72% aqueous distillate and left 10–18% charred residue. After the pyrolysis, the residue was found to consist of some water-soluble materials, in addition to char and undecomposed cellulose.

The hemicelluloses undergo thermal decomposition very readily. The hemicelluloses reacted more readily than cellulose during heating. The thermal degradation of hemicelluloses begins above 373 K during heating for 48 h; hemicelluloses and lignin are depolymerized by steaming at high temperature for a short time. The methoxyl content of wet meals decreased at 493 K. The stronger effect of the heating rate on the formation of biochar from biomass than from coal may be attributed to the cellulose content of the biomass (Demirbas 2006). It is well known that the heating rate has a significant effect on the pyrolysis of cellulose. The quick devolatilization of the biomass in rapid pyrolysis favors the formation of char with high porosity and high reactivity. The decreased formation of char at the higher heating rate was accompanied by an increased formation of tar. The net effect is a decrease in the volatile fuel production and an increased yield of biochar cellulose converted to levoglucosan at temperatures above 535 K (Freudenberg and Neish 1968).

2.4.2 Adhesives from Biomass

Wood adhesives play an essential role in the industry. Main wood valorization technologies include pulp and paper making, bio-oil by pyrolysis, synthesis gas by gasification, sugar by hydrolysis, ethanol by sugar fermentation, and finally adhesives by alkali liquefaction and polymerization. Lignin is a complex, high molecular weight polymer built of hydroxyphenylpropane units. There are two types of phenolic resins: resol and novalac. A phenolic resin for partially replacing phenol was used with modified organosolv lignin in phenol-formaldehyde (PF) resin production. Organosolv lignin-phenol-formaldehyde (LPF) resins were produced in a two-step preparation with different additions of lignin.

Adhesion is the state in which two surfaces are held together by interfacial forces, which may be valence forces, interlocking action, or both. The adhesives used in commercial wood composite products are usually synthetic polymer resins, based on the condensation reaction of formaldehyde with phenol, urea, resorcinol or melamine (Cetin and Ozmen 2003). Some structural, semi-structural and non-structural wood adhesives are listed in Table 2.8. Approximately 1 million metric

Table 2.8 Structural, semi-structural, and non-structural wood adhesives

Structural	Phenol-formaldehyde
	Resorcinol-formaldehyde
	Phenol-resorcinol-formaldehyde
	Emulsion polymer/iso-cyanate
	Melamine-formaldehyde
	Melamine-urea-formaldehyde
	Epoxy
	Iso-cyanate
	Urea-formaldehyde
	Casein
Semi-structural	Cross-linked polyvinyl acetate
	Polyurethane
Non-structural	Polyvinyl acetate
	Animal
	Soybean
	Elastomeric construction
	Elastomeric contact
	Starch

tons of urea-formaldehyde resins are produced annually. More than 70% of this urea-formaldehyde resin is used by the forest products industry for a variety of purposes.

Fermentation residues were obtained by growing the anaerobic cellulolytic bacteria *Ruminococcus albus 7* or *Clostridium thermocellum* ATCC 27405 on a fibrous fraction derived from lucerne (*Medicago sativa L.*). These were converted to an effective co-adhesive for phenol-formaldehyde (PF) bonding of aspen veneer sheets to one another. Testing of the resulting plywood panels revealed that the adhesive was formulated to contain 30% of its total dry weight as fermentation residue (Weimer et al. 2005).

The wood adhesive market is very large, and there are significant problems due to volatile organic compounds and toxic chemicals in many adhesives. In addition, most of the adhesives are derived from depleting petrochemical resources. An environmentally friendly wood adhesive based on renewable resources and produced by microbial fermentation has been explored (Haag et al. 2006).

A method is described for making adhesives from biomass. The liquefaction oil is prepared from lignin-bearing plant material and a phenolic fraction is extracted. The phenolic fraction is reacted with formaldehyde to yield a phenol-formaldehyde resin. At present, the production of wood composites mainly relies on the petrochemical-based and formaldehyde-based adhesives such as PF resins and urea-formaldehyde (UF) resins (Liu and Li 2007). Phenol-formaldehyde adhesives are used to manufacture plywood, flakeboard, and fiberglass insulation. Phenolic resins occur only in dark, opaque colors and can therefore be used only to manufacture dark-colored products.

Phenolic resins are some of the principal thermofixed synthetic polymers, the third most important polymeric matrix for composites (Lubin 1969), and are also known for their high temperature resistance (Kopf and Little 1991). Pure phenolic resin can be obtained through the condensation reaction between phenol (C_6H_5OH) and formaldehyde (CH_2O), producing methylene bridges between the phenol molecules (Leite et al. 2004). Thermosetting polymers make excellent structural adhesives because they undergo irreversible chemical change, and on reheating, they do not soften and flow again. They form cross-linked polymers that have high strength, have resistance to moisture and other chemicals, and are rigid enough to support high, long-term static loads without deforming. Phenolic, resorcinolic, melamine, iso-cyanate, urea, and epoxy are examples of types of wood adhesives that are based on thermosetting polymers.

Resistance to chemical attack is generally improved by resin impregnation, which protects the underlying wood and reduces movement of liquid into the wood. Resistance to acids can be obtained by impregnating with phenolic resin, and to alkalis, by impregnating with furfural resin. The adhesion properties of different natural fillers without the addition of coupling agents by considering the different filler morphology were investigated. The adhesion behavior has been determined in a qualitative way from microscopic observation and has been quantified; its influence on mechanical properties decreases, since in this type of material, tensile strength usually decreases (Crespo et al. 2007).

The reaction of formaldehyde with lignin model compounds in acidic medium is shown to give fast cross-linking of alkali-substituted phenolic and etherified phenolic lignin model compounds at *meta*-positions to the aromatic hydroxyl groups. This reaction differs from the reaction of formaldehyde with phenolic lignin model compounds in alkaline conditions, where the reaction with formaldehyde always occurs at position *ortho-/para-* to the aromatic hydroxyl group (Demirbas and Ucan 1991).

The lignin degradation products and their sodium salts can be converted into very weak organic acids by treating mineral acids. Most phenols have equilibrium constants (K_a) in the neighborhood of 10^{-10}, and are thus considerably weaker than the carboxylic acids (K_a about 10^{-5}). Most phenols are weaker than carbonic acid ($CO_2 + H_2O$), and hence, unlike carboxylic acids, do not dissolve in aqueous bicarbonate solutions. Indeed, phenols are conveniently liberated from their salts by action of carbonic acid (K_a about 10^{-7}) (Morrison and Boyd 1983).

Organosolv lignin-phenol-formaldehyde (LPF) resins were produced in a two-step preparation with different additions of lignin. The method selected for the manufacture of lignin resins dealt with modification of the lignin by the methylolation route. Organosolv lignin-based resins showed comparatively good strength and stiffness. The tensile strength properties of test samples made from organosolv lignin resins were equal to or better than those of test samples made from PF resin only (Cetin and Ozmen 2003).

2.4.3 Valorization of Wood

Forests are principal global economic resources, as well as ecological resources. Forests have played a big role in the development of human societies. The prime direct or marketable product of most forests is wood for use as timber, fuelwood, pulp and paper, providing some 3.4 billion cubic meters of timber-equivalent a year globally. Asia and Africa use 75% of global wood fuels.

The lumber, plywood, pulp, and paper industries burn their own wood residues in large furnaces and boilers to supply 60% of the energy needed to run factories. The availability of fuelwood from the forest is continually declining at an ever-increasing rate due to indiscriminate deforestation and slow regeneration, as well as afforestation. The fuelwoods generally used by local people were identified and analyzed quantitatively to select a few species with the best fuelwood characteristics so that plantation on non-agricultural lands could be undertaken (Jain 1992).

The reduction of particle size and moisture content, together with the most appropriate storage and handling systems are necessary for an efficiently operated wood waste combustion system. Size reduction may be carried out in several stages in a hog or attrition mill, with screening before and in between. The moisture in residues may be reduced either by mechanical pressing, air drying, or the use of hot air dryers, or a combination of all three. Generally slabs, edgings, peeler cores, veneer waste and trimmings would be transported by mechanical conveyors.

Forest energy involves the use of forest biomass, which is currently not being used in the traditional forest products industries. Essentially, this means the forest residues are left after forest harvesting, including residual trees and scrub or under-managed woodland. Forest residues alone count for some 50% of the total forest biomass and are currently left in the forest to rot.

The principal sources of waste wood are two waste streams: municipal solid waste (MSW) and construction and demolition waste (C&DW). Municipal solid waste (MSW) is waste from residential, commercial, institutional, and industrial sources. Each generates distinctly different types of wood waste, with differing degrees and levels of recyclability. The primary components of wood waste are used lumber, shipping pallets, trees, branches, and other wood debris from construction and demolition clearing and grubbing activities. Construction and demolition waste is defined as solid, largely inert waste resulting from the construction, repair, demolition, or razing of buildings, roads and other structures. The term also includes debris from the clearing of land for construction. Other sources of waste wood include chemically treated wood from railroad ties, telephone and utility poles, miner poles, crosstie, constructors, pier and dock timbers, untreated wood from logging and silvicultural operations, and industrial waste wood outside the MSW and C&DW streams. Chemical treatments and costs of collection make much of this material difficult to recover.

The wood waste preparation process generally involves hogging, dewatering, screening, size reduction, bulk storage, blending and drying prior to combustion so as to ensure a reliable and consistent supply of quality fuel to the burners. The handling, treatment and storage of wood waste fuel are considerably more costly and troublesome than that required for traditional fossil fuels.

The most limiting factors for using wood waste as fuel for power generation are transportation costs and its energy content. The economics of wood waste energy generation become more attractive as traditional fuel prices increase.

2.5 Biofuels from Biomass

2.5.1 Introduction

The term biofuel is referred to as solid, liquid, or gaseous fuels that are predominantly produced from biorenewable or combustible renewable feedstocks. Liquid biofuels are important in the future because they replace petroleum fuels. The biggest difference between biofuels and petroleum feedstocks is oxygen content. Biofuels are non-polluting, locally available, accessible, sustainable and reliable fuel obtained from renewable sources. Electricity generation from these fuels is a promising method to use in the near future. The future of biomass electricity generation lies in biomass-integrated gasification/gas turbine technology, which offers high-energy conversion efficiencies.

Biofuels can be classified based on their production technologies: first-generation biofuels (FGBs), second-generation biofuels (SGBs), third-generation biofuels (TGBs), and fourth-generation biofuels.

The FGBs refer to biofuels made from sugar, starch, vegetable oils, or animal fats using conventional technology. The basic feedstocks for the production of first-generation biofuels are often seeds or grains such as wheat, which yields starch that is fermented into bioethanol, or sunflower seeds, which are pressed to yield vegetable oil that can be used in biodiesel. Table 2.9 shows the classification of renewable biofuels based on their production technologies.

Second and third-generation biofuels are also called advanced biofuels, and are made from non-food crops, wheat straw, corn, wood, energy crops using advanced technology. Algae fuel, also called *oilgae* or third-generation biofuel, is a biofuel from algae. Algae are low-input/high-yield (30 times more energy per acre than land) feedstocks to produce biofuels using more advanced technology. On the other hand, an appearing fourth-generation is based on the conversion of vegoil and biodiesel into biogasoline using the most advanced technology.

Table 2.9 Classification of renewable biofuels based on their production technologies

Generation	Feedstock	Example
First-generation biofuels	Sugar, starch, vegetable oils, or animal fats	Bioalcohols, vegetable oil, biodiesel, biosyngas, biogas
Second-generation biofuels	Non-food crops, wheat straw, corn, wood, solid waste, energy crop	Bioalcohols, bio-oil, bio-DMF, biohydrogen, bio-Fischer–Tropsch diesel, wood diesel
Third-generation biofuels	Algae	Vegetable oil, biodiesel
Fourth-generation biofuels	Vegetable oil, biodiesel	Biogasoline

There are two global biorenewable liquid transportation fuels that might replace gasoline and diesel fuel. These are bioethanol and biodiesel. Bioethanol is a good alternate fuel that is produced almost entirely from food crops. Biodiesel has become more attractive recently because of its environmental benefits.

Transport is one of the main energy-consuming sectors. It is assumed that biodiesel can be used as a fossil diesel replacement and that bioethanol can be used as a gasoline replacement. Biomass-based energy sources for heat, electricity, and transportation fuels are potentially carbon dioxide neutral and recycle the same carbon atoms. Due to the widespread availability of biorenewables, this fuel technology may potentially employ more people than fossil fuel-based technology.

Renewable liquid biofuels for transportation have recently attracted much attention in different countries all over the world because of their renewability, sustainability, common availability, regional development, rural manufacturing jobs, reduction of greenhouse gas emissions, and its biodegradability.

2.5.2 *Bioethanol*

Bioethanol can be used directly in cars designed to run on pure ethanol or blended with gasoline to make "gasohol." Anhydrous ethanol is required for blending with gasoline. No engine modification is typically needed to use the blend. Ethanol can be used as an octane-boosting, pollution-reducing additive in unleaded gasoline.

Figure 2.4 shows the world production of ethanol and biodiesel between 1980 and 2007 (Demirbas 2008). Between 1991 and 2001, world ethanol production rose

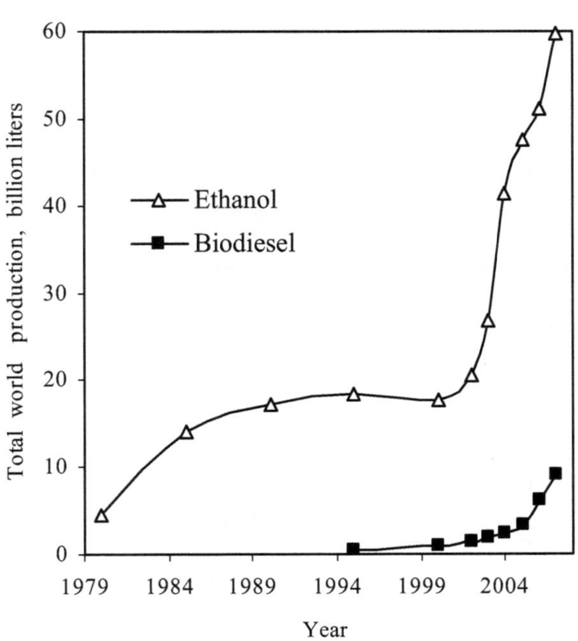

Fig. 2.4 World production of ethanol and biodiesel (1980–2007)

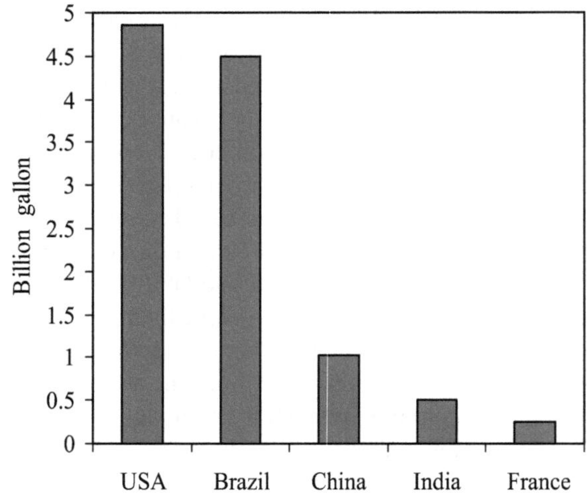

Fig. 2.5 The top five bioethanol producers (billion gallons) in 2006

from around 16 billion liters a year to 18.5 billion liters. From 2001 to 2007, production is expected to have tripled, to almost 60 billion liters a year. Brazil was the world's leading ethanol producer until 2005 when US production roughly equaled Brazil's. The US became the world's leading ethanol producer in 2006. China holds a distant but important third place in world rankings, followed by India, France, Germany and Spain. Figure 2.5 shows the top five bioethanol producers in 2006 (RFA 2007).

The continued increases in the price of crude oil in 2005 and 2006 resulted in a reversal of the traditional relationship between the price of biomass energy and that of crude oil, something not seen since the 1930s. As a consequence of the high prices of traded crude oil, many countries advanced their biofuel goals and, in the case of Brazil and the US, large production gains occurred. Table 2.10 shows the world production of ethanol during 2004 and 2007.

Table 2.10 World production of ethanol during 2004 and 2007 (billion gallons)

Country	2004	2005	2006	2007	Growth rate 2006–2007	Share of total 2007
United States	3.40	3.90	4.86	6.49	33.5%	38.3%
Brazil	3.87	4.25	4.71	5.96	26.5%	35.2%
China	0.93	0.93	0.94	0.99	5.3%	5.9%
India	0.33	0.29	0.44	0.65	47.7%	3.8%
France	0.22	0.24	0.24	0.30	25.0%	1.8%
Others	2.00	2.10	2.40	2.55	6.3%	15.0%
Total	10.75	11.71	13.59	16.94	24.6%	100.0%

2.5.3 Biodiesel

The other significant biofuel is biodiesel, which is currently produced from vegetable oils, animal fats and grease by transesterification. The vegetable oils with carbon chain lengths of between 16 and 22 carbon atoms are generally in the form of triacylglycerides (TAG), which on transesterification with methanol produce fatty acid methyl ester (FAME) as the precursor to biodiesel and glycerol as a byproduct. Vegetable oil (m)ethyl esters, commonly referred to as biodiesel, are prominent candidates as alternative diesel fuels. The name biodiesel has been given to transesterified vegetable oil to describe its use as a diesel fuel. After FAME purification and testing for compliance with either EN 14214 or ASTM D6751 standards the product can be sold as biodiesel and used as blends – typically B5 (5% biodiesel) to B20, depending on the engine warranties.

There has been renewed interest in the use of vegetable oils for making biodiesel due to its less polluting and renewable nature as compared to conventional diesel, which is a fossil fuel leading to a potential exhaustion. Biodiesel is technically competitive with or offers technical advantages compared to conventional petroleum diesel fuel. The vegetable oils can be converted to their (m)ethyl esters via the transesterification process in the presence of a catalyst. Methyl, ethyl, 2-propyl and butyl esters can be prepared from vegetable oils through transesterification using potassium and/or sodium alkoxides as catalysts. The purpose of the transesterification process is to lower the viscosity of the oil. Ideally, transesterification is potentially a less expensive way of transforming the large, branched molecular structure of the bio-oils into smaller, straight-chain molecules of the type required in regular diesel combustion engines.

The biodiesel esters were characterized for their physical and fuel properties including density, viscosity, iodine value, acid value, cloud point, pure point, gross heat of combustion and volatility. The biodiesel fuels produced slightly lower power and torque, and higher fuel consumption than No. 2 diesel fuel. Biodiesel is better than diesel fuel in terms of sulfur content, flash point, aromatic content and biodegradability.

The cost of biodiesels varies depending on the base stock, geographic area, variability in crop production from season to season, the price of the crude petroleum and other factors. Biodiesel has over double the price of petroleum diesel. The high price of biodiesel is in large part due to the high price of the feedstock. However, biodiesel can be made from other feedstocks, including beef tallow, pork lard, and yellow grease.

The commercial resource base for vegetable oils comprises about 20 different species with soybean oil, rapeseed (Colza), palm/palm kernel oil, sunflower, and coconut oils being the largest sources. Most of the biodiesel that is currently made uses soybean oil, methanol, and an alkaline catalyst. The high value of soybean oil as a food product makes production of a cost-effective fuel very challenging. However, there are large amounts of low-cost oils and fats such as restaurant waste and animal fats that could be converted to biodiesel. The problem with processing these

Table 2.11 Biodiesel production (thousand tons)

	2004	2005	2006
Germany	1035	1669	2681
France	348	492	775
Italy	320	396	857
Malaysia		260	600
United States	83	250	826
Czech Republic	60	133	203
Poland		100	150
Austria	57	85	134
Slovakia	15	78	89
Spain	13	73	224
Denmark	70	71	81
UK	9	51	445
Other EU	6	36	430
Total	2016	3694	7495

low-cost oils and fats is that they often contain large amounts of free fatty acids (FFA) that cannot be converted to biodiesel using an alkaline catalyst.

Biodiesel is an environmentally friendly alternative liquid fuel that can be used in any diesel engine without modification. If the biodiesel valorized efficiently, it would be beneficial in terms of the environment and the local population, job creation, provision of modern energy carriers to rural communities, avoiding urban migration, and reduction of CO_2 and sulfur levels in the atmosphere. Table 2.11 shows the biodiesel production between 2004 and 2006.

The second generation biofuels include renewable and green diesels. The former is a technology that incorporates vegetable oils into the crude oil-derived diesel production process to produce a renewable carbon-based diesel with no oxygen content and a very high cetane number, while the latter is the production of middle distillates by means of Fischer–Tropsch catalysts, using synthesis gas produced by the gasification of biomass. Fischer–Tropsch-like catalysts (synthol process) can also produce ethanol and mixed alcohols.

2.5.4 Bio-Oil

The term bio-oil is used mainly to refer to liquid fuels. There are several reasons for bio-oils to be considered as relevant technologies by both developing and industrialized countries. They include energy security reasons, environmental concerns, foreign exchange savings, and socioeconomic issues related to the rural sector.

Bio-oils are liquid or gaseous fuels made from biomass materials, such as agricultural crops, municipal wastes and agricultural and forestry byproducts via biochemical or thermochemical processes. They can serve as substitutes for conventional fuels in vehicle engines, either totally or partially in a blend.

Biomass is dried and then converted to oily product known as bio-oil, by very quick exposure to heated particles in a fluidized bed. The char and gases produced are combusted to supply heat to the reactor, while the product oils are cooled and condensed. The bio-oil is then shipped by truck from these locations to the hydrogen production facility. It is more economical to produce bio-oil at remote locations and then ship the oil, since the energy density of bio-oil is higher than biomass. For this analysis, it was assumed that the bio-oil would be produced at several smaller plants, which are closer to the sources of biomass, such that lower-cost feedstocks can be obtained.

The purpose of this analysis was to assess the economic feasibility of producing hydrogen from biomass via two thermochemical processes: (1) gasification followed by reforming of the syngas (H_2+CO), and (2) fast pyrolysis followed by reforming of the carbohydrate fraction of the bio-oil. In each process, the water–gas shift is used to convert the reformed gas into hydrogen, and pressure swing adsorption is used to purify the product.

Pyrolysis (cracking) is defined as the cleavage to smaller molecules by thermal energy. Hydrogen can be produced economically from woody biomass. Biomass can be thermally processed through gasification or pyrolysis to produce hydrogen. The main gaseous products from biomass are the following:

$$\text{Pyrolysis of biomass} \rightarrow H_2 + CO_2 + CO + \text{Gaseous and liquid hydrocarbons} \quad (2.7)$$

$$\text{Catalytic steam reforming of biomass} \rightarrow H_2 + CO_2 + CO \quad (2.8)$$

$$\text{Fischer–Tropsch synthesis of } (H_2 + CO) \rightarrow \text{Gaseous and liquid hydrocarbons} \quad (2.9)$$

The conventional pyrolysis of biomass is associated with the product of interest that is the high charcoal continent, but the fast pyrolysis is associated with tar, at low temperatures (675–775 K), and/or gas, at high temperatures.

2.5.5 Biogas

The organic fraction of almost any form of biomass, including sewage sludge, animal wastes and industrial effluents, can be broken down through anaerobic digestion into a methane and carbon dioxide mixture known as *biogas*. Biogas is an environment friendly, clean, cheap and versatile fuel.

Anaerobic digestion of biowastes occurs in the absence of air, the resulting gas, biogas, is a mixture consisting mainly of methane and carbon dioxide. Biogas is a valuable fuel, which is produced in digesters filled with the feedstock like dung or sewage. The digestion is allowed to continue for a period from 10 days to a few weeks.

Biogas can be obtained from several sources. It is obtained from decomposing organic material. Biogas is composed of methane (CH_4), carbon dioxide (CO_2), air, ammonia, carbon monoxide, hydrogen, sulfur gases, nitrogen and oxygen. Among its components, methane is the most important one, particularly for the combustion process in vehicle engines. CH_4 and CO_2 make up around 90% of the gas volume produced. The main constituents of landfill gas are methane and carbon dioxide, both of which are major contributors to global warming. Because of the widely varying nature of the contents of landfill sites the constituents of landfill gases vary widely.

Total solids (TS) content in the slurry is determined by drying it in an oven at 378 K until a constant weight is obtained. The dried solid samples from the TS determination are ignited at 1225 K in a furnace for 7 min. The loss in weight is taken as the volatile solids of the substrate slurry.

After the first 6 days of digestion, methane production from manure increases exponentially, after 16 days it reaches a plateau value, and at the end of the 20th day, the digestion reaches the stationary phase. For the wheat straw and the mixtures of manure/straw the rates of digestion are lower than that of manure.

The maximum daily biogas productions are between 4 and 6 days. During a 30-day digestion period, around 80–85% of the biogas is produced in the first 15–18 days. This implies that the digester retention time can be designed to 15–18 days instead of 30 days.

For first 3 days, the methane yield is almost 0% and carbon dioxide generation is almost 100%. In this period, digestion occurs as fermentation to carbon dioxide. The yields of methane and carbon dioxide gases are fifty-fifty on day 11. At the end of day 20, the digestion reaches the stationary phase. The methane content of the biogas is in the range of 73–79% for the runs, the remainder being principally carbon dioxide. During digestion, the volatile fatty acid concentration is lower and the pH higher. The pH of the slurry with manure increased from 6.4 initially, to 6.9–7.0 at the maximum methane production rate. The pH of the slurry with wheat straw is around 7.0–7.1 at the maximum methane production rate.

2.5.6 Fischer–Tropsch Liquids

Syngas (a mixture of carbon monoxide and hydrogen) produced by gasification of fossil fuels or biomass can be converted into a large number of organic compounds that are useful as chemical feedstocks, fuels, and solvents. Many of the conversion technologies were developed for coal gasification but process economics have resulted in a shift to natural gas-derived syngas. These conversion technologies successively apply similarly to biomass-derived biosyngas. Franz Fischer and Hans Tropsch first studied conversion of syngas into larger, useful organic compounds in 1923 (Spath and Mann 2000).

Biofuels, including Fischer–Tropsch liquids, are important as they affect a wide array of topics, including energy security, environmental concerns, foreign exchange savings, and socioeconomic issues related to the rural sector. Catalytic conversion will be a primary tool for industry to produce valuable fuels, chemicals, and materials from biomass platform chemicals. The catalytic conversion of biomass is best developed for synthesis gas or syngas. Economic considerations dictate that the current production of liquid fuels from syngas must make use of natural gas as the hydrocarbon source. Biomass is the only renewable that can meet our demand for carbon-based liquid fuels and chemicals. Biofuels as well as green motor fuels produced from biomass by Fischer–Tropsch synthesis (FTS) are the most modern biomass-based transportation fuels. Green motor fuels are the renewable replacement to petroleum-based diesel. Biomass energy conversion facilities are important for obtaining bio-oil by pyrolysis. The aim of FTS is the synthesis of long-chain hydrocarbons from a CO and H_2 gas mixture. The products from FTS are mainly aliphatic straight-chain hydrocarbons (C_xH_y). In addition to the C_xH_y, branched hydrocarbons, unsaturated hydrocarbons, and primary alcohols are formed in minor quantities. The FTS process is a process capable of producing liquid hydrocarbon fuels from biosyngas. The large hydrocarbons can be hydrocracked to form mainly diesel of excellent quality. The process for producing liquid fuels from biomass, which integrates biomass gasification with FTS, converts a renewable feedstock into a clean fuel.

Fischer–Tropsch synthesis (FTS) produces mainly straight-chain hydrocarbons from a syngas rich in CO and H_2, and catalysts are usually employed. Typical operating conditions for FTS are temperatures of 475–625 K and very high pressure, depending on the desired products. The product range includes light hydrocarbons such as methane (CH_4) and ethane (C_2H_6), propane (C_3H_8) butane (C_4H_{10}), gasoline (C_5–C_{12}), diesel (C_{13}–C_{22}), and light and heavy waxes (C_{23}–C_{33}). The distribution of the products depends on the catalyst and the process conditions (temperature, pressure, and residence time). The synthesis gas must have very low tar and particulate matter content.

The FTS is described by the following set of equations (Schulz 1999):

$$nCO + (n + m/2)\, H_2 \rightarrow C_nH_m + nH_2O \qquad (2.10)$$

where n is the average length of the hydrocarbon chain and m is the number of hydrogen atoms per carbon. All reactions are exothermic and the product is a mixture of different hydrocarbons, where paraffin and olefins are the main parts.

The first FTS plants began operation in Germany in 1938 but closed down after World War II. Then in 1955, Sasol, a world-leader in the commercial production of liquid fuels and chemicals from coal and crude oil, started Sasol I in Sasolburg, South Africa. Following the success of Sasol I, Sasol II and III, located in Secunda, South Africa, came on line in 1980 and 1982, respectively. The FTS is an essential step to convert carbon-containing feedstocks to liquid fuels such as diesel. Major advantages of the FTS are: (1) flexibility in feedstocks (natural gas, coal, biomass), (2) large and even sustainable resources, (3) ultraclean (low sulfur content) products, (4) suitability for converting difficult to process resources. A major drawback of the Fischer–Tropsch process however is the polymerization-like nature of the process, which yields a wide spectrum of products, ranging from compounds with low molecular mass like methane, to products with very high molecular mass like heavy waxes.

2.5.7 Biohydrogen

Hydrogen is not a primary fuel. It must be manufactured from water with either fossil or non-fossil energy sources. Widespread use of hydrogen as an energy source could improve global climate change, energy efficiency, and air quality. The thermochemical conversion processes, such as pyrolysis, gasification and steam gasification are available for converting the biomass to a more useful energy. The yield from steam gasification increases with increasing water-to-sample ratio. The yields of hydrogen from the pyrolysis and the steam gasification increase with increasing temperature. The list of some biomass material used for hydrogen production is given in Table 2.12. Hydrogen-powered fuel cells are an important enabling technology for the hydrogen future and more-efficient alternatives to the combustion of gasoline and other fossil fuels. Hydrogen has the potential to solve two major energy problems: reducing dependence on petroleum and reducing pollution and greenhouse gas emissions.

A fuel cell is a device or an electrochemical engine that converts the energy of a fuel directly to electricity and heat without combustion. Fuel cells consist of two electrodes sandwiched around an electrolyte. When oxygen passes over one electrode and hydrogen over the other, electricity is generated. Fuel cells running on hydrogen derived from a renewable source would emit nothing but water vapor. Fuel cells are clean, quiet, and efficient.

Hydrogen is currently more expensive than conventional energy sources. There are different technologies presently being practiced to produce hydrogen economically from biomass. Hydrogen can be produced by pyrolysis from biomass. It can be burned to produce heat or passed through a fuel cell to produce electricity. Biomass represents a large potential feedstock resource for environmentally clean hydrogen production. It lends itself to both biological and thermal conversion processes. In the thermal path hydrogen can be produced in two ways: direct gasification and pyrolysis to produce liquid bio-oil, followed by steam reforming. Hydrogen can

Table 2.12 List of some biomass material used for hydrogen production

Biomass species	Main conversion process
Bionut shell	Steam gasification
Olive husk	Pyrolysis
Tea waste	Pyrolysis
Crop straw	Pyrolysis
Black liquor	Steam gasification
Municipal solid waste	Supercritical water extraction
Crop grain residue	Supercritical fluid extraction
Pulp and paper waste	Microbial fermentation
Petroleum basis plastic waste	Supercritical fluid extraction
Manure slurry	Microbial fermentation

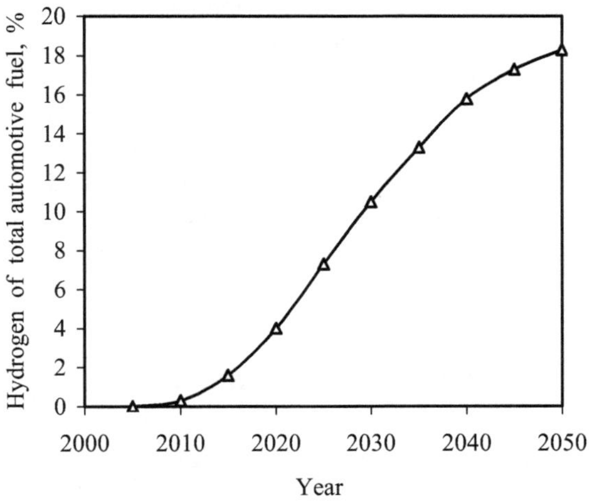

Fig. 2.6 Share of hydrogen in total automotive fuel in the future

alternately be produced by supercritical water gasification of biomass (Byrd et al. 2007).

Biohydrogen technology will play a major role in the future because it can utilize renewable sources of energy. Figure 2.6 shows the share of hydrogen in total automotive fuel consumption in the world as a futuristic view (Demirbas 2008).

Biological generation of hydrogen (biohydrogen) technologies provide a wide range of approaches to generate hydrogen, including direct biophotolysis, indirect biophotolysis, photofermentations, and dark fermentation. Biological hydrogen production processes are found to be more environmentally friendly and less energy intensive as compared to thermochemical and electrochemical processes. Researchers have investigated hydrogen production with anaerobic bacteria since the 1980s.

There are three types of microorganisms of hydrogen generation: cyanobacteria, anaerobic bacteria, and fermentative bacteria. The cyanobacteria directly decompose water into hydrogen and oxygen in the presence of light by photosynthesis. Photosynthetic bacteria use organic substrates like organic acids. Anaerobic bacteria use organic substances as the sole source of electrons and energy, converting them into hydrogen. Biohydrogen can be generated using bacteria such as *Clostridia* by temperature, pH control, reactor hydraulic retention time (HRT) and other factors of the treatment system.

Biological hydrogen can be generated from plants by biophotolysis of water using microalgae (green algae and cyanobacteria), fermentation of organic compounds, and photodecomposition of organic compounds by photosynthetic bacteria. To produce hydrogen by fermentation of biomass, a continuous process using a non-sterile substrate with a readily available mixed microflora is needed (Hussy et al. 2005). A successful biological conversion of biomass to hydrogen depends strongly on the processing of raw materials to produce feedstock, which can be fermented by the microorganisms.

Hydrogen production from the bacterial fermentation of sugars has been examined in a variety of reactor systems. Hexose concentration has a greater effect on

H_2yields than the hydraulic retention time (HRT). Flocculation is also an important factor in the performance of the reactor.

Hydrogen gas is a product of the mixed acid fermentation of *Escherichia coli*, the butylene glycol fermentation of *Aerobacter*, and the butyric acid fermentations of *Clostridium* spp. Anaerobic fermentation was conducted to improve hydrogen fermentation of food waste in a leaching-bed reactor by heat-shocked anaerobic sludge, and also to investigate the effect of dilution rate on the production of hydrogen and metabolites in hydrogen fermentation.

2.5.8 Sugar from Biomass

Cellulose is a remarkable pure organic polymer, consisting solely of units of anhydroglucose held together in a giant straight-chain molecule. Cellulose must be hydrolyzed to glucose before fermentation to ethanol. Conversion efficiencies of cellulose to glucose may be dependent on the extent of chemical and mechanical pretreatments to structurally and chemically alter the pulp and paper mill wastes. The method of pulping, the type of wood, and the use of recycled pulp and paper products also could influence the accessibility of cellulose to cellulase enzymes. Hemicelluloses (arabinoglycuronoxylan and galactoglucomannans) are related to plant gums in composition, and occur in much shorter molecule chains than cellulose. The hemicelluloses, which are present in deciduous woods chiefly as pentosans and in coniferous woods almost entirely as hexosanes, undergo thermal decomposition very readily. Hemicelluloses are derived mainly from chains of pentose sugars, and act as the cement material holding together the cellulose micelles and fiber. Cellulose is insoluble in most solvents and has a low accessibility to acid and enzymatic hydrolysis. Hemicelluloses are largely soluble in alkalis and, as such, are more easily hydrolyzed.

Hydrolysis (saccharification) breaks down the hydrogen bonds in the hemicellulose and cellulose fractions into their sugar components: pentoses and hexoses. These sugars can then be fermented into bioethanol.

Cellulose (hexosan, $(C_6H_{10}O_5)_n$) hydrolysis produces glucose (a hexose, $C_6H_{12}O_6$). The hydrolysis of cellulose is catalyzed by mineral acids and enzymes.

Hexosan to hexose:

$$(C_6H_{10}O_5)_n + nH_2O \rightarrow nC_6H_{12}O_6 \tag{2.11}$$

Hemicelluloses hydrolysis produces both hexose and pentose sugars: mannose, galactose, xylose and arabinose that are not all fermented with existing strains.

Pentosan to pentose:

$$(C_5H_8O_4)_m + mH_2O \rightarrow mC_5H_{10}O_5 \tag{2.12}$$

The most commonly applied methods can be classified into two groups: chemical hydrolysis (dilute and concentrated acid hydrolysis) and enzymatic hydrolysis. In addition, there are some other hydrolysis methods in which no chemicals or enzymes are applied. For instance, lignocellulose may be hydrolyzed by gamma ray or electron beam irradiation, or microwave irradiation. However, those processes are commercially unimportant.

Both enzymatic and chemical hydrolyses require a pretreatment to increase the susceptibility of cellulosic materials. In the chemical hydrolysis, the pretreatment and the hydrolysis may be carried out in a single step. There are two basic types of acid hydrolysis processes commonly used: dilute acid and concentrated acid, each with variations.

The dilute acid process is conducted under high temperature and pressure, and has a reaction time in the range of seconds or minutes, which facilitates continuous processing. As an example, using a dilute acid process with 1% sulfuric acid in a continuous flow reactor at a residence time of 0.22 min and a temperature of 510 K with pure cellulose, provides a yield over 50% sugars. In this case, 1000 kg of dry wood would yield about 164 kg of pure ethanol. The combination of acid and high temperature and pressure dictate special reactor materials, which can make the reactor expensive. The first reaction converts the cellulosic materials to sugar and the second reaction converts the sugars to other chemicals. Unfortunately, the conditions that cause the first reaction to occur are also the right conditions for the second to occur.

The biggest advantage of dilute acid processes is their fast rate of reaction, which facilitates continuous processing. Since five-carbon sugars degrade more rapidly than six-carbon sugars, one way to decrease sugar degradation is to have a two-stage process. The first stage is conducted under mild process conditions to recover the five-carbon sugars, while the second stage is conducted under harsher conditions to recover the six-carbon sugars.

Hydrolysis of cellulosic materials by concentrated sulfuric or hydrochloric acids is a relatively old process. The concentrated acid process uses relatively mild temperatures, and the only pressures involved are those created by pumping materials from vessel to vessel. Reaction times are typically much longer than for dilute acid. This method generally uses concentrated sulfuric acid followed by a dilution with water to dissolve and hydrolyze or convert the substrate into sugar. This process provides a complete and rapid conversion of cellulose to glucose and hemicelluloses to five-carbon sugars with little degradation. The critical factors needed to make this process economically viable are to optimize sugar recovery and cost effectively recovers the acid for recycling. The solid residue from the first stage is dewatered and soaked in a 30 to 40% concentration of sulfuric acid for 1 to 4 h as a precellulose hydrolysis step. The solution is again dewatered and dried, increasing the acid concentration to about 70%. After reacting in another vessel for 1 to 4 h at low temperatures, the contents are separated to recover the sugar and acid. The sugar/acid solution from the second stage is recycled to the first stage to provide the acid for the first stage hydrolysis.

The primary advantage of the concentrated acid process is the potential for high sugar recovery efficiency. The acid and sugar are separated via ion exchange and then the acid is reconcentrated via multiple effect evaporators. The low temperatures and pressures employed allow the use of relatively low-cost materials such as fiberglass tanks and piping. The low temperatures and pressures also minimize the degradation of sugars. Unfortunately, it is a relatively slow process and cost effective acid recovery systems have been difficult to develop. Without acid recovery, large quantities of lime must be used to neutralize the acid in the sugar solution. This neutralization forms large quantities of calcium sulfate, which requires disposal and creates additional expense.

Another basic method of hydrolysis is enzymatic hydrolysis. Enzymes are naturally occurring plant proteins that cause certain chemical reactions to occur. There are two technological developments: enzymatic and direct microbial conversion methods.

The chemical pretreatment of the cellulosic biomass is necessary before enzymatic hydrolysis. The first application of enzymatic hydrolysis was used in separate hydrolysis and fermentation steps. Enzymatic hydrolysis is accomplished by cellulolytic enzymes. Different kinds of "cellulases" may be used to cleave the cellulose and hemicelluloses. A mixture of endoglucanases, exoglucanases, β-glucosidases and cellobiohydrolases is commonly used. The endoglucanases randomly attack cellulose chains to produce polysaccharides of shorter length, whereas exoglucanases attach to the non-reducing ends of these shorter chains and remove cellobiose moieties. β-glucosidases hydrolyze cellobiose and other oligosaccharides to glucose (Demirbas 2001).

For enzymes to work efficiently, they must obtain access to the molecules to be hydrolyzed. This requires some kind of pretreatment process to remove hemicelluloses and break down the crystalline structure of the cellulose or removal of the lignin to expose hemicelluloses and cellulose molecules.

2.6 Barriers to the Development of Biofuels

First-generation biofuels (FGBs) refer to biofuels made from sugar, starch, vegetable oils, or animal fats using conventional technology. FGBs produced from food crops such as grains, sugar beet and oilseeds are limited in their ability to achieve targets for oil-product substitution, climate change mitigation, and economic growth. A possible exception that appears to meet many of the acceptable criteria is bioethanol produced from sugar cane. The basic feedstocks for the production of first-generation biofuels are often seeds or grains such as wheat, which yields starch that is fermented into bioethanol, or sunflower seeds, which are pressed to yield vegetable oil that can be used in biodiesel.

The production of FGBs such as sugar cane ethanol in Brazil, corn ethanol in US, oilseed rape biodiesel in Germany, and palm oil biodiesel in Malaysia are characterized by mature commercial markets and well-understood technologies. Future targets and investment plans suggest strong growth will continue in the near future.

Second-generation biofuels (SGBs) produced from (larger) feedstocks from lignocellulosic materials include cereal straw, forest residues, bagasse, and purpose-grown energy crops such as vegetative grasses and short-rotation forests. The SGBs could avoid many of the concerns facing FGBs and potentially offer greater cost reduction potential in the longer term. Many of problems associated with FGBs can be addressed by the production of biofuels manufactured from agricultural and forest residues and from non-food crop feedstocks. Low-cost crop and forest, wood process wastes, and the organic fraction of municipal solid wastes can all be used as lignocellulosic feedstocks.

The key messages related to FGBs and SGBs are as follows:

1. Technical barriers remain for second biofuel production.
2. Production costs are uncertain and vary with the feedstock available.
3. There is no clear candidate for "best technology pathway" between the competing biochemical and thermochemical routes.
4. Even at high oil prices, SGBs will probably not become fully commercial nor enter the market for several years to come without significant additional government support.
5. Considerably more investment in research, development, demonstration and deployment is needed to ensure that future production of the various biomass feedstocks can be undertaken sustainably and that the preferred conversion technologies, including those more advanced but only at the research and development stage, are identified and proven to be viable.
6. Once proven, there will be a steady transition from FGBs (with the exception of sugarcane ethanol that will continue to be produced sustainably in several countries).

There are some barriers to the development of biofuel production. They are technological, economical, supply, storage, safety, and policy barriers. Reducing these barriers is one of the driving factors in the government's involvement in biofuel and biofuel research and development. The production of biofuels from lignocellulosic feedstocks can be achieved through two very different processing routes: biochemical and thermochemical. Technical barriers for enzymatic hydrolysis include: low specific activity of current commercial enzymes, high cost of enzyme production, and lack of understanding of enzyme biochemistry and mechanistic fundamentals.

References

Alanne, K., Sari, A. 2006. Distributed energy generation and sustainable development. Renew Sust Energy Rev 10:539–558.

Arkesteijn, K., Oerlemans, L.A.G. 2005. The early adoption of green power by Dutch households. An empirical exploration of factors influencing the early adoption of green electricity for domestic purposes. Energy Policy 33(2):183–196.

Babu, B.V., Chaurasia, A.S. 2003. Modeling for pyrolysis of solid particle: kinetics and heat transfer effects. Energy Convers Manage 44:2251–2275.

Bushnell, D.J., Haluzok, C., Dadkhah-Nikoo, A. 1989. Biomass fuel characterization: testing and evaluating the combustion characteristics of selected biomass fuels. Bonneville Power Administration, Corvallis, OR.

Byrd, A.J., Pant, K.K., Gupta, R.B. 2007. Hydrogen production from glucose using Ru/Al_2O_3 catalyst in supercritical water. Ind Eng Chem Res 46:3574–3579.

Cetin, N.S., Ozmen, N. 2003. Studies on lignin-based adhesives for particleboard panels. Turk J Agric For 27:183–189.
Crespo, J.E., Balart, R., Sanchez, L., Lopez, J. 2007. Mechanical behaviour of vinyl plastisols with cellulosic fillers. Analysis of the interface between particles and matrices. Int J Adhesion Adhesives 27:422–428.
Demirbas, A. 1991. Fatty and resin acids recovered from spruce wood by supercritical acetone extraction. Holzforschung 45:337–339.
Demirbas, A. 1997. Calculation of higher heating values of biomass fuels. Fuel 76:431–434.
Demirbas, A. 1998. Determination of combustion heat of fuels by using non-calorimetric experimental data. Energy Edu Sci Technol 1:7–12.
Demirbas, A. 2000. Mechanisms of liquefaction and pyrolysis reactions of biomass. Energy Convers Manage 41:633–646.
Demirbas, A. 2001. Biomass resource facilities and biomass conversion processing for fuel and chemicals. Energy Convers Manage 42:1357–1378.
Demirbas, A. 2002. Biodiesel from vegetable oils via transesterification in supercritical methanol. Energy Convers Manage 43:2349–2356.
Demirbas. 2004. Combustion characteristics of different biomass fuels. Prog Energy Combus Sci 30:219–230.
Demirbas, A. 2006. Production and characterization of bio-chars from biomass via pyrolysis. Energy Sources Part A 28:413–422.
Demirbas, A. 2008. Biodiesel: a realistic fuel alternative for diesel engines. Springer, London.
Demirbas, A., Ucan, H.I. 1991. Low temperature pyrolysis of black liquor and polymerization of products in alkali aqueous medium. Fuel Sci Technol Int 9:93–105.
Demirbas, A., Gullu, D., Caglar, A., Akdeniz, F. 1997. Determination of calorific values of fuel from lignocellulosics. Energy Sources 19:765–770.
Domburg, G., Rossinskaya, G., Sergseva, V. 1974. Study of thermal stability of b-ether bonds in lignin and its models. In: Proceedings of 4th International Conference Thermal Analysis, Budapest, vol. 2, p. 221.
Elliott, D. 1999. Prospects for renewable energy and green energy markets in the UK. Renewable Energy 16:1268–1271.
Fengel, D., Wegener, G. 1983. Wood: chemistry, ultrastructure, reactions, chap. 7, p. 326. Walter de Gruyter, Berlin.
Freudenberg, K., Neish, A.C. 1968. Constitution and biosynthesis of lignin. Springer, Berlin Heidelberg New York.
Fridleifsson, I.B. 2003. Status of geothermal energy amongst the world's energy sources. Geothermics 32:379–388.
Glasser, W.G., Sarkanen, S. (Eds.) 1989. Lignin: properties and materials. American Chemical Society, Washington, DC.
Goldemberg, J., Coelho, S.T. 2004. Renewable energy: traditional biomass vs. modern biomass. Energy Policy 32:711–714.
Haag, A.P., Geesey, G.G., Mittleman, M.W. 2006. Bacterially derived wood adhesive. Int J Adhesion Adhesives 26:177–183.
Hamelinck, C., Faaij, A. 2006. Outlook for advanced biofuels. Energy Policy 34(17):3268–3283.
Hussy, I., Hawkes, F.R., Dinsdale, R., Hawkes, D.L. 2005. Continuous fermentative hydrogen production from sucrose and sugarbeet. Int J Hydrogen Energy 30:471–483.
IPCC (Intergovernmental Panel on Climate Change). 2007. IPCC Climate Change: the Fourth Assessment Report (AR4) of the United Nations Intergovernmental Panel on Climate Change (IPCC). UN, Paris.
Jain, R.K. 1992. Fuelwood characteristics of certain hardwood and softwood tree species of India Biores Technol 41:129–133.
Karaosmanoglu, F., Aksoy, H.A. 1988. The phase separation problem of gasoline-ethanol mixture as motor fuel alternatives. J Thermal Sci Technol 11:49–52.
Khan, A.A., de Jong, W., Jansens, P.J., Spliethoff, H. 2009. Biomass combustion in fluidized bed boilers: potential problems and remedies. Fuel Proc Technol 90:21–50.

Kopf, P.W., Little, A.D. 1991. Phenolic Resins. In: Kirk-Othmer Encyclopedia of Chemical Technology, vol. 18. Wiley-Interscience, New York.

Kuznetsov, S.A., Kuznetsov, B.N., Aleksandrova, N.B., Danlov, V.G., Zhizhaev, A.M. 2005. Obtaining arabinigalactan, dihydrate quercetin and microcrystalline cellulose using mechanochemical activation. Chem Sustain Develop 13:261–268.

Larson, E.D. 1993. Technology for fuels and electricity from biomass. Annu Rev Energy Environ 18:567–630.

Leite, J.L., Pires, A.T.N., Ulson de Souza, S.M.A.G., Ulson de Souza, A.A. 2004. Characterisation of a phenolic resin and sugar cane pulp composite. Brazilian J Chem Eng 21:253–260.

Liu, Y., Li, K. 2007. Development and characterization of adhesives from soy protein for bonding wood. Int J Adhesion Adhesives 27:59–67.

Lubin, G. 1969. Handbook of fiberglass and advanced plastics composites. Van Nostrand Reinhold, New York.

Morrison, R.T., Boyd, R.N. 1983. Organic chemistry, 4th edn. Allynn and Bacon, New York.

Murphya, H., Niitsuma, H. 1999. Strategies for compensating for higher costs of geothermal electricity with environmental benefits. Geothermics 28:693–711.

Peterson, A.A., Vogel, F., Lachance, R.P., Froling, M., Antal, M.J., Tester, J.W. 2008. Thermochemical biofuel production in hydrothermal media: a review of sub- and supercritical water technologies. Energy Environ Sci 1:32–65.

Ragland, K.W., Aerts, D.J., Baker, A.J. 1991. Properties of wood for combustion analysis. Biores Technol 37:161–168.

Reddy, S.S., Kotaıah, B., Reddy, N.S.P., Velu, M. 2006. The removal of composite reactive dye from dyeing unit effluent using sewage sludge derived activated carbon. Turk J Eng Env Sci 30:367–373.

RFA (Renewable Fuels Association). 2007. Ethanol Industry Statistics. RFA, Washington, DC.

Sarkanen, K.V., Ludwig, C.H. 1971. Lignins: occurrence, formation, structure and reactions. Wiley, New York.

Schulz, H. 1999. Short history and present trends of FT synthesis. Appl Catal A General 186:1–16.

Shafizadeh, F. 1982. Introduction to pyrolysis of biomass. J Anal Appl Pyrolysis 3:283–305.

Shafizadeh, F. 1985. In: Fundamentals of thermochemicals biomass conversion, Overend, R.P., Milne, T.A., Mudge, L.K. (Eds.) Elsevier Applied Science, London.

Sheehan, J., Dunahay, T., Benemann, J., Roessler, P. 1998. A look back at the US Department of Energy's Aquatic Species Program: biodiesel from algae. (NREL) Report: NREL/TP-580-24190. National Renewable Energy Laboratory, Golden, CO.

Spath, P.L., Mann, M.K. 2000. Life cycle assessment of hydrogen production via natural gas steam reforming. TP-570-27637, November. National Renewable Energy Laboratory, Golden, CO.

Theander, O. 1985. In: Fundamentals of thermochemical biomass conversion, Overend, R.P., Milne, T.A., Mudge, L.K. (Eds.) Elsevier Applied Science, London.

Tillman, D.A. 1978. Wood as an energy resource. Academic, New York.

UNDP (United Nations Development Programme). 2000. World Energy Assessment. Energy and the challenge of sustainability. UNDP, New York.

Viswanathan, B. 2006. An introduction to energy sources. Indian Institute of Technology, Madras, India.

WEC (World Energy Council). 2007. Survey of energy resources. World Energy Council, London.

Weimer, P.J., Koegel, R.G., Lorenz, L.F., Frihart, C.R., Kenealy, W.R. 2005. Wood adhesives prepared from lucerne fiber fermentation residues of Ruminococcus albus and Clostridium thermocellum. Appl Microbiol Biotechnol 66:635–640.

Zanzi, R. 2001. Pyrolysis of biomass. Dissertation, Royal Institute of Technology, Department of Chemical Engineering and Technology, Stockholm.

Chapter 3
Biorefinery

3.1 Introduction

The need for energy is continuously increasing, due to increases in industrialization and population. The growth of the world's energy demand raises urgent problems. The larger part of petroleum and natural gas reserves is located within a small group of countries. For example, the Middle East countries have 63% of the global reserves and are the dominant supplier of petroleum. This energy system is unsustainable because of equity issues as well as environmental, economic, and geopolitical concerns that have far reaching implications. Interestingly, the renewable energy resources are more evenly distributed than fossil or nuclear resources. Also, the energy flows from renewable resources are more than three orders of magnitude higher than the current global energy need.

Bioenergy is one of the most important components to reduce greenhouse gas emissions and act as a substitute to fossil fuels (Goldemberg 2000; Dincer 2008). Renewable energy is one of the most efficient ways to achieve sustainable development.

Plants use photosynthesis to convert solar energy into chemical energy. It is stored in the form of oils, carbohydrates, proteins, etc. This plant energy is converted to biofuels. Hence, biofuels are primarily a form of solar energy. For biofuels to succeed at replacing large quantities of petroleum fuel, the feedstock availability needs to be as high as possible. There is an urgent need to design integrated biorefineries that are capable of producing transportation fuels and chemicals.

In recent years, recovery of the liquid transportation biofuels from biorenewable feedstocks has become a promising method for the future. The biggest difference between biorenewable and petroleum feedstocks is oxygen content. Biorenewables have oxygen levels from 10 to 45%, while petroleum has essentially none, making the chemical properties of biorenewables very different from petroleum (Demirbas 2008; Balat 2008, 2009). For example, biorenewable products are often more polar and some easily entrain water and can therefore be acidic.

Table 3.1 Availability of modern transportation fuels

Fuel type	Availability	
	Current	Future
Gasoline	Excellent	Moderate-poor
Bioethanol	Moderate	Excellent
Biodiesel	Moderate	Excellent
Compressed natural gas (CNG)	Excellent	Moderate
Hydrogen for fuel cells	Poor	Excellent

There are two global transportation fuels: gasoline and diesel fuel. The main transportation fuels that can be obtained from biomass using various processes are: sugar ethanol, cellulosic ethanol, grain ethanol, biodiesel, pyrolysis liquids, green diesel, green gasoline, butanol, methanol, syngas liquids, biohydrogen, algae diesel, algae jet fuel, and hydrocarbons. The use of renewable liquid biofuels for transportation has recently attracted attention in different countries because of their renewability, sustainability, common availability, regional development, potential manufacturing jobs in rural communities, reduction of greenhouse gas emissions, and their biodegradability (Demirbas 2008). Table 3.1 shows the availability of modern transportation fuels. Transportation fuels and petroleum-fueled, biorenewable-fueled (ReFueled), and biorenewable electricity (ReElectricity) powered vehicles are shown in Fig. 3.1.

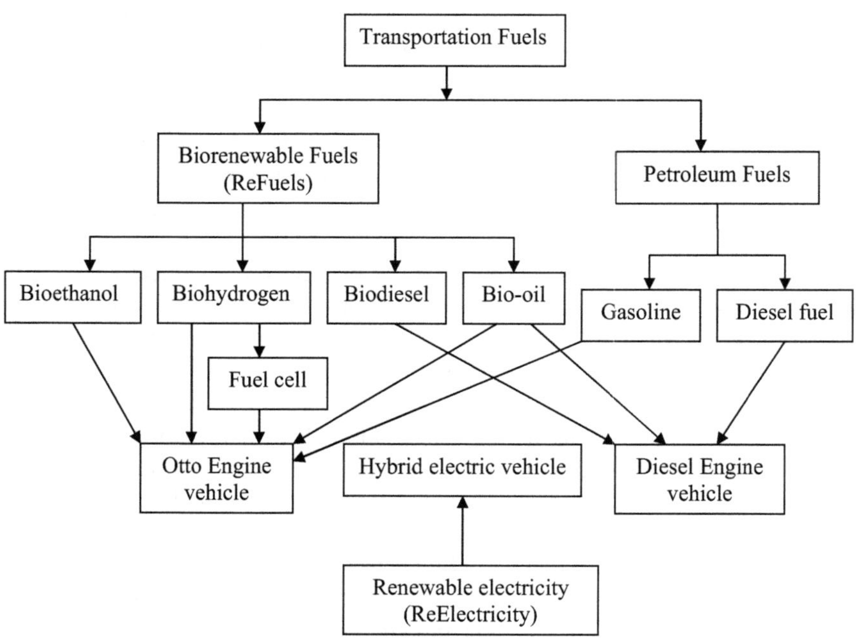

Fig. 3.1 Transportation fuels and petroleum-fueled, biorenewable-fueled (ReFuel), and biorenewable electricity (ReElectricity) powered vehicles

The term *biofuel* or *biorenewable fuel* (ReFuel) refers to a solid, liquid, or gaseous fuel that is predominantly produced from biomass. Liquid biofuels that are considered for use in different parts of the world fall into the following categories: (1) bioalcohols, (2) vegetable oils and biodiesels, and (3) biocrude and synthetic oils. Biofuels are important because they replace petroleum fuels. It is expected that the demand for biofuels will rise in the future. Biofuels are substitute fuel sources to petroleum; however, some still include a small amount of petroleum in the mixture. With an increasing number of governments now supporting this cause in the form of mandates and other policy initiatives, the biofuel industry is poised to grow at a phenomenal rate (Balat 2007; Demirbas 2002, 2003, 2007; Demirbas and Karslioglu 2007; Khoiyangbam 2008; Chhetri and Islam 2008).

Policy drivers for biorenewable liquid biofuels have attracted support for rural development and economic opportunities for developing countries (Keskin 2009). The European Union is ranked third in terms of biofuel production worldwide, behind Brazil and the US. In Europe, Germany is the largest, and France the second largest producer of biofuels.

The term modern biomass is generally used to describe the biomass used through efficient and clean combustion technologies and a sustained supply of biomass resources, environmentally sound and competitive fuels, heat and electricity using modern conversion technologies. Biomass, as an energy source, has two striking characteristics. Firstly, biomass is the only renewable organic resource, and is also one of the most abundant resources. Secondly, biomass fixes carbon dioxide in the atmosphere by photosynthesis. Biomass thermochemical conversion technologies such as pyrolysis and gasification are certainly not the most important options at present, since combustion is responsible for over 97% of the world's bioenergy production. Ethanol and fatty acid (m)ethylester (biodiesel) as well as diesel produced from biomass by Fischer–Tropsch synthesis (FTS) are modern biomass-based transportation fuels. Liquid transportation fuels can be economically produced by biomass-integrated gasification-Fischer–Tropsch (BIG-FT) processes. Modern biomass produced in a sustainable way excludes traditional uses of biomass as fuelwood and includes electricity generation and heat production, as well as transportation fuels, from agricultural and forest residues and solid waste. On the other hand, traditional biomass is produced in an unsustainable way and it is used as a non-commercial source – usually with very low efficiencies for cooking in many countries. Biomass energy potentials and current use in different regions are given in Table 3.2 (Parikka 2004).

Bio-based products are prepared for economic use by an optimal combination of different methods and processes (physical, chemical, biological, and thermal). It is therefore necessary that basic biorefinery concepts are developed. A biorefinery, as a new approach, is a processing plant where biomass feedstocks are converted and extracted into a spectrum of valuable products. Biorefining refers to fractionating biomass into various separated products that possibly undergo further chemical, biochemical, biological, and thermochemical processing and separation. The molecules produced by biorefining can be obtained using thermal, chemical, mechanical, enzymatic or microbial processes, and can be used in transport fuels, therapeutics, food additives or as secondary chemicals with a range of applications.

Table 3.2 Biomass energy potentials and current use in different regions

Biomass potential	North America	Latin America	Asia	Africa	Europe	Middle East	Former USSR	World
Woody biomass	12.8	5.9	7.7	5.4	4.0	0.4	5.4	41.6
Energy crops	4.1	12.1	1.1	13.9	2.6	0.0	3.6	37.4
Straw	2.2	1.7	9.9	0.9	1.6	0.2	0.7	17.2
Other	0.8	1.8	2.9	1.2	0.7	0.1	0.3	7.6
Potential sum (EJ/yr)	19.9	21.5	21.4	21.4	8.9	0.7	10.0	103.8
Use (EJ/yr)	3.1	2.6	23.2	8.3	2.0	0.0	0.5	39.7
Use/Potential (%)	16.0	12.0	108.0	39.0	22.0	7.0	5.0	38.0

By means of co-producing chemicals the production costs of secondary energy carriers (fuels, heat, power) potentially could become more profitable, especially when biorefining is integrated into the existing chemical, material and power industries. The co-production of bioproducts, materials, chemicals, transportation fuels, power and/or heat in fully optimized integrated biorefinery systems will be required.

3.2 Definitions of Biorefinery

A biorefinery is a facility that integrates biomass conversion processes and equipment to produce fuels, power, and value-added chemicals from biomass. The biorefinery concept is analogous to today's crude oil refineries, which produce multiple fuels and products from petroleum. *Biorefinery* refers to the conversion of biomass feedstock into a host of valuable chemicals and energy with minimal waste and emissions. In a broad definition biorefineries process all kinds of biomass (all organic residues, energy crops, and aquatic biomass) into numerous products (fuels, chemicals, power and heat, materials, and food and feed). Figure 3.2 shows a schematic diagram of the biorefinery concept.

A biorefinery is a conceptual model for future biofuel production where both fuels and high-value co-product materials are produced. Biorefineries would simultaneously produce biofuels as well as bio-based chemicals, heat, and power. Biorefineries would present more economical options, where bio-based chemicals are co-products of liquid fuel. Future biorefineries would be able to mimic the energy efficiency of modern oil refining through extensive heat integration and co-product development. Heat that is released from some processes within the biorefinery could be used to meet the heat requirements for other processes in the system (WI 2007).

Reliable designs of equipment for the thermochemical stages of biomass conversion have yet to be widely demonstrated or shown to be able to continuously pro-

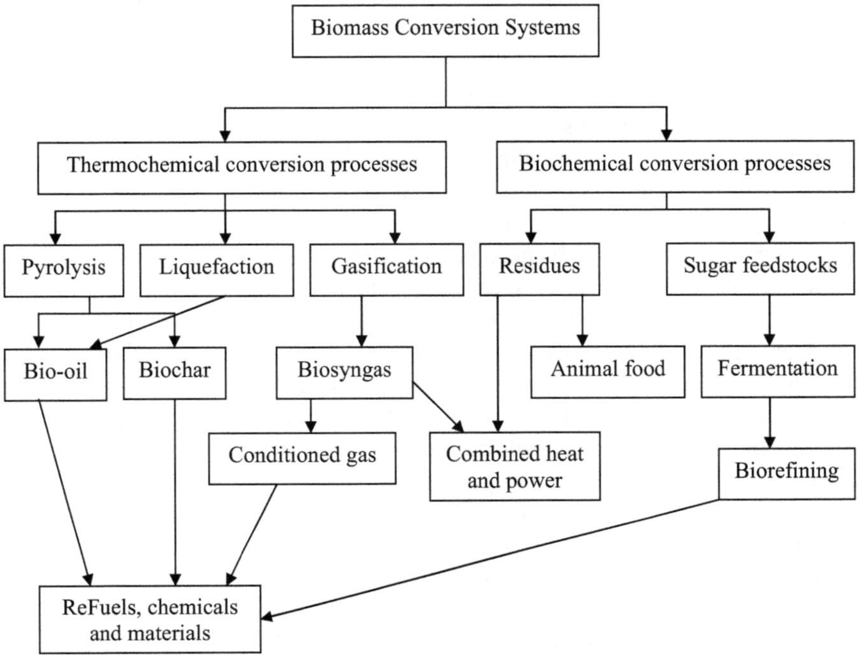

Fig. 3.2 Schematic diagram of biorefinery concept

duce synthesis gas of the required quality. Synthesis gas has to be free of nitrogen; this requires the use of oxygen (an expensive process), a pyrolytic process optimized for gas, or a multi-stage process. The main non-technical barriers to acceptance of liquid biofuels, especially in the transport sector, are related to the costs of production, available markets, taxation policies and legislation as well as blending and distribution. From an economic point of view the use of agricultural crops, without subsidy, is too expensive to produce either bioethanol or biodiesel at a price competitive with untaxed petrol or diesel fuel, while processes for using lower cost lignocellulosic materials have not been perfected. There are a number of pilot-scale and demonstration plants either operating, under development, or planned, with fully commercial developments not expected until the next decade or two (GBEP 2007).

Developed and developing countries clearly have different goals in development of biomass energy. Biomass energy is promoted in developed countries to displace fossil fuels, particularly in the transportation sector, whereas its use serves as basic survival purposes in developing countries. The combination of higher-value products with higher volume energy production and employing any number of conversion technologies has the greatest potential for making fuels, chemicals and materials, and power from biomass competitive. Obtaining modern biofuels, biopower, and bioproducts from biomass can be realized only in integrated biorefineries. Biorefineries too will use only those technology platforms that are the most cost effective for converting a certain type of biomass into a certain collection of desired end products. This section reviews the current biorenewable fuel valorization facilities as well as the future importance of biorefineries. The development of biorefinery technologies is important and these technologies are also very promising.

The term biorefinery is widely thought of as concept of multi-products from varying biomass feedstocks. A biorefinery processes biomass into value-added product streams. In theory, anything that uses biomass and makes more than one product is a biorefinery. Biorefinery is analogous to that of a petroleum refinery processing a range of crude oils. This very simple definition captures a wide range of existing, emerging, and advanced process concepts. Examples of existing biorefineries include corn processors and pulp and paper mills.

The concept is analogous to a combined use of fluid catalytic cracking, thermal cracking, and hydrocracking technology to convert the higher boiling range fractions of crude oil into more useful lower boiling range products. Only a few petroleum refineries use all the available conversion technologies; this will be the same for biorefineries, which will only utilize the technologies that are the most cost effective for converting one type of biomass into certain end products. For example, for crops and agricultural waste, it would be better to convert the biomass to bio-oil near the farm and transporting the high-density bio-oil to a central facility for processing rather than transporting the low-density biomass.

The biorefinery concept attempts to apply the methods that have been applied to the refining of petroleum to biomass conversion. The goal is to maximize the value of the products obtained from the biomass. The goal of the integrated biorefinery program area is to support the establishment of integrated biorefineries through partnerships with industry and academia.

3.2.1 Main Technical and Non-Technical Gaps and Barriers to Biorefineries

There are many technical and non-technical gaps and barriers related to the implementation and commercialization of the biorefinery. Current technical barriers with the use of energy crops are associated with the cost of production and difficulties in harvesting and storing the material grown, especially for annual or other crops that have to be harvested within a narrow time period in the autumn. Transportation costs are of prime importance when calculating the overall costs of biomass; hence, local or regional production of biomass is most favorable. Other technical problems associated with growing energy crops include provision of nutrients and control of pests and disease.

The major non-technical barriers are restrictions or prior claims on use of land (food, energy, amenity use, housing, commerce, industry, leisure or designations as areas of natural beauty, special scientific interest, etc.), as well as the environmental and ecological effects of large areas of monoculture. For example, vegetable oils are a renewable and potentially inexhaustible source of energy with energy content close to diesel fuel. On the other hand, extensive use of vegetable oils may cause other significant problems such as starvation in developing countries. Vegetable oil fuels are not acceptable because they are more expensive than petroleum fuels.

There are few technical barriers to building biomass-fired facilities at any scale, from domestic uses to around 50 MW, above which considerations of the availability and cost of providing fuel become significant. In general however, the capacity and generating efficiency of biomass plants are considerably less than those of modern natural gas-fired turbine systems. The main non-technical limitations to investment in larger systems are economic, or in some countries the problems reflect planning conditions and public opinion, where a clear distinction may not be made between a modern effective biomass energy plant and an older polluting incinerator design.

In addition to the technical challenges of commercializing advanced biorefineries, there are also large infrastructure barriers. These barriers are associated with the development of new agricultural infrastructure for the collection and storage of crop wastes. An integrated feedstock supply system must be developed that can supply the feedstock needs in a sustainable fashion at a reasonable cost. Infrastructure issues could be as significant as the technical issues when considering overall production costs.

3.3 The History of Biorefinery

The process of converting biorenewables into useful and higher-value compounds is not new. For example, sugarcane has been used in the production of bioethanol since 6000 BC. Lactic acid was first discovered by C.W. Scheele in 1780. Its production was developed into an industrial-level fermentation by A. Boehringer in Germany in 1895, with subsequent uses in the food, leather, and textile industries. Even to this day, lactic acid is used as a preservative, flavor enhancer, and aciduant in the food industry.

Ethanol has been used by humans since prehistory as the intoxicating ingredient of alcoholic beverages. The ancient Egyptians produced alcohol by naturally fermenting vegetative materials. Also in ancient times, the Chinese discovered the art of distillation, which increases the concentration of alcohol in fermented solutions. Dried residues were discovered on 9000-year-old pottery found in China. Its isolation as a relatively pure compound was first achieved by Persian alchemist, Zakariya Razi. Ethanol was first prepared synthetically in 1826 through the independent efforts of Henry Hennel in the UK. Ethanol was used as lamp fuel in the US as early as 1840. In the 1970s, however, the ethanol industry began to reemerge when ethanol was used as a fuel extender during gasoline shortages caused by the OPEC oil embargoes.

Ethanol was used in Germany and France as early as 1894 by the then incipient industry of internal combustion engines. Brazil has utilized ethanol as a fuel since 1925. By that time, the production of ethanol was 70 times greater than the production and consumption of petrol.

The emergence of transesterification can be dated back to as early as 1846 when Rochieder described glycerol preparation through ethanolysis of castor oil. Since that time alcoholysis has been studied in many parts of the world. Transesterification of a vegetable oil was conducted as early as 1853, by scientists E. Duffy and J. Patrick, many years before the first diesel engine became functional. Life for the diesel engine began in 1893 when the famous German inventor Dr. Rudolph Diesel published a paper entitled "The Theory and Construction of a Rational Heat Engine." What the paper described was a revolutionary engine in which air would be compressed by a piston to a very high pressure thereby causing a high temperature. Dr. Diesel designed the original diesel engine to run on vegetable oil.

Dr. Diesel's prime model, a single 3-m iron cylinder with a flywheel at its base, ran on its own power for the first time in Augsburg, Germany on August 10, 1893 (in remembrance of this event, August 10th has been declared International Biodiesel Day). This engine stood as an example of Diesel's vision because it was powered by peanut oil, a biofuel, though not strictly biodiesel, since it was not transesterified. He believed that the utilization of a biomass fuel was the real future of his engine. In a 1912 speech, Dr. Diesel said, "the use of vegetable oils for engine fuels may seem insignificant today, but such oils may become, in the course of time, as important as petroleum and the coal-tar products of the present time."

The use of vegetable oils as alternatives to petroleum was proposed in the beginning of the 1980s. The advantages of vegetable oils as diesel fuel are: liquid nature and portability, readily available, renewability, higher heat content (about 88% of No. 2 petroleum diesel fuel), lower sulfur content, lower aromatic content, biodegradability. The energy supply concerns of the 1970s renewed interest in biodiesel, but commercial production did not begin until the late 1990s.

Since the 1980s, biodiesel plants have opened in many European countries, and some cities have run buses on biodiesel, or blends of petrol and biodiesels. More recently, Renault and Peugeot have approved the use of biodiesel in some of their truck engines. Recent environmental and domestic economic concerns have prompted a resurgence in the use of biodiesel throughout the world. In 1991, The European Community (EC) proposed a 90% tax deduction for the use of biofuels, including biodiesel. Biodiesel plants are now being built by several companies in Europe; each of these plants will produce up to 1.5 million gallons of fuel per year. The European Union accounted for nearly 89% of all biodiesel production worldwide in 2005.

A historical and technological outline of industrial resources includes aspects on the beginnings of sugar production, starch hydrolysis, wood saccharification, formation and separation of furfural, cellulose and pulp production, synthesis of levulinic acid, lipids, vanillin from lignin, and lactic acid fermentation. During the industrial revolution, coal became the preferred energy source, displacing biomass fuels such as wood. Since then a steady migration towards fossil fuels has continued, moving further away from biomass, not only for energy but also for sources of chemicals used to make everyday items. An excellent example of this is furfural, which can be obtained from oat hulls, and used in the production of nylon. Some current aspects of biorefinery research and development since the beginning of the 1990s reveal that integrated processes and biorefinery technology have become objects of research and development.

Many of the currently used bio-based industrial products are results of direct physical or chemical treatment and processing of biomass: cellulose, starch, oil, protein, lignin, and terpenes. Also, due to the help of biotechnological processes and methods, feedstock chemicals are produced, such as ethanol, butanol, acetone, lactic and itaconic acid, as well as amino acids.

The topic of biorefineries as a means of processing industrial material and efficient utilization of renewable products is well known and applied worldwide, in almost every developed and emerging country. This tendency is motivated by the rising cost of oil and the need to move away from petrochemical-based systems. The global impact of biomass refining includes aspects of the fossil fuel carbon-processing industries, the existing bio-based economy. The sustainability of integrated biorefining systems depends on aspects of the modeling and analysis of the agriculture/forestry ecosystem, and also on technical and economic considerations underlining the importance of the biorefinery and the growing request for new products.

Furthermore, the technology of sugar refining, starch production, and oil milling, the separation of proteins as feed, and the extraction of chlorophyll for industrial use with alfalfa as raw material were of great historical importance. Processes like wet grinding of crops and the production of ethanol as well as acetic, lactic, and citric acid were once fundamental in the nineteenth and twentieth century. Figure 3.3 shows the global trends in biofuel production by region.

Today's biorefinery technologies are based on (a) the utilization of the whole plant or complex biomass, and (b) on the integration of traditional and modern processes for utilization of biological raw materials (Kamm et al. 2005). In the nineteenth and the beginning of the twentieth century, large-scale utilization of renewable resources focused on pulp and paper production from wood, saccharification

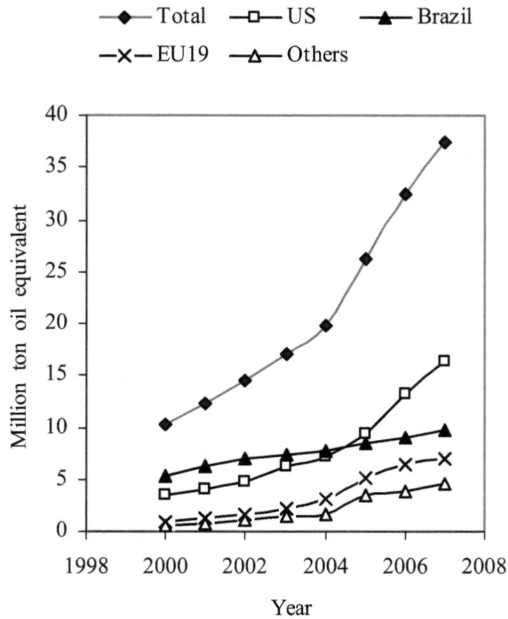

Fig. 3.3 Global trends in biofuels production by region

Table 3.3 Biorefinery demonstration projects: plant capacity, planned or under-construction

Name	Location	Participants	Annual capacity (million gallons)
Iogen	Shelley, Idaho, Canada	Iogen Corporation, Goldman Sachs, Royal Dutch Shell Oil Company, and others	18 (in first plant) 250 (in future plants) Cellulosic ethanol
BlueFire Ethanol	Lancaster, California, USA	BlueFire Ethanol Fuels, Inc.	32 Cellulosic ethanol
Poet	Sioux Falls, South Dakota, Ohio	Broin Companies	65 Corn ethanol
Ecofin, LLC	Washington County, Kentucky	University of Cincinnati, University of Kentucky, Alltech Inc.	1.3 Corncob ethanol
ICM, Incorporated	St. Joseph, Missouri	AGCO Eng; NCAUR-ARS-Peoria; CERES Inc.; Edens. Sys. Corp.; Novozymes NA Inc.; Sun EtOH Inc.; Others	1.5 Lignocellulosic ethanol
Lignol Innovations	Commerce City, Colorado	Canada's Lignol Energy Corporation	2.5 Cellulosic ethanol
Royal Nedalco	Rotterdam Area, Netherlands	Ethanol Reach Association	60 Cellulosic ethanol
Mascoma	Rome, New York	Mascoma Corporation, DuPont–Danisco	0.5 Cellulosic ethanol
Pacific Ethanol	Boardman, Oregon	Pacific Ethanol, Inc.	2.7 Lignocellulosic ethanol
Abengoa Bioenergy	Colwich, Kansas	Abengoa Bioenergy LLL	25 Lignocellulosic ethanol
Verenium	Jenning, Louisana	Verenium Corporation	1.5 Cellulosic ethanol
RSE Pulp	Old Town, Maine	RSE Pulp & Chemical LLC Company	2.3 Cellulosic ethanol
Flambeau	Park Falls, Wisconsin	Flambeau LLC	14.2 Lignocellulosic ethanol
NewPage	Wisconsin Rapids, Wisconsin	NewPage Corporation	12.8 Biofuels, FT liquids
Choren	Freiberg, Germany	Choren Industries	1.5 Biofuels, FT liquids
Alico	LaBelle, Florida	Alico Inc., Bioeng. Res. Inc., Fayetteville, Arkansas, Washington Group Int. of Boise, Idaho, Emmaus Found. Inc., and others	7.5 Ethanol, electricity, ammonia, hydrogen

Table 3.3 (*continued*) Biorefinery demonstration projects: plant capacity, planned or under-construction

Name	Location	Participants	Annual capacity (million gallons)
Range Fuels	Broomfield, Colorado	Range Fuels Inc.	1.2 Biofuels, FT liquids
Flambeau River	Jennings, Louisiana	Flambeau River BioFuels	6.5 Biofuels, FT liquids

of wood, nitration of cellulose for guncotton and viscose silk, as well as the production of soluble cellulose for fibers, fat curing, and the production of furfural for nylon (see Table 3.3). Advanced biorefineries are envisioned to serve as the foundation of a new bioindustry. By exploiting new chemical, biological and mechanical technologies, they offer the promise of greatly expanding the use of renewable plant-based materials, as well as a means of transitioning to a more energy efficient and environmentally sustainable chemical and energy economy.

3.4 Petroleum Refinery and Biorefinery

In 1858, Dr. Abraham Gesner, a Canadian physician and amateur geologist, developed and patented the extraction of a lamp fuel from asphalt rock, which he named kerosene. At that time, kerosene, which was an extremely high-value lighting fuel for lamps, and was the primary product of petroleum refining. For a while, distillation of kerosene for lamps was the mainstay of the new petroleum industry. Gasoline was merely a byproduct of kerosene production from crude oil and, until the early 1900s, there was no significant demand for it. The first petrochemical, aside from carbon black manufactured on an industrial scale, was iso-propyl alcohol, by Standard Oil of New Jersey in 1920.

When a mixture of two liquids of different boiling points is heated to its boiling point, the vapor contains a higher mole fraction of the liquid with the lower boiling point than the original liquid, i. e., the vapor is enriched in the more volatile component. If this vapor is now condensed, the resultant liquid will also been enriched in the more volatile component. This is the principle of batch fractional distillation, and in a distillation column many such cycles are performed continuously, allowing almost complete separation of liquid components. A generalized distillation column is shown in Fig. 3.4. The first step in the refining of crude oil, whether in a simple or a complex refinery, is the separation of the crude oil into fractions (fractionation or fractional distillation). These fractions are mixtures containing hydrocarbon compounds whose boiling points lie within a specified range.

Crude oil is a complex mixture that is between 50 and 95% hydrocarbon by weight. The first step in refining crude oil involves separating the oil into different hydrocarbon fractions by distillation. An oil refinery cleans and separates the

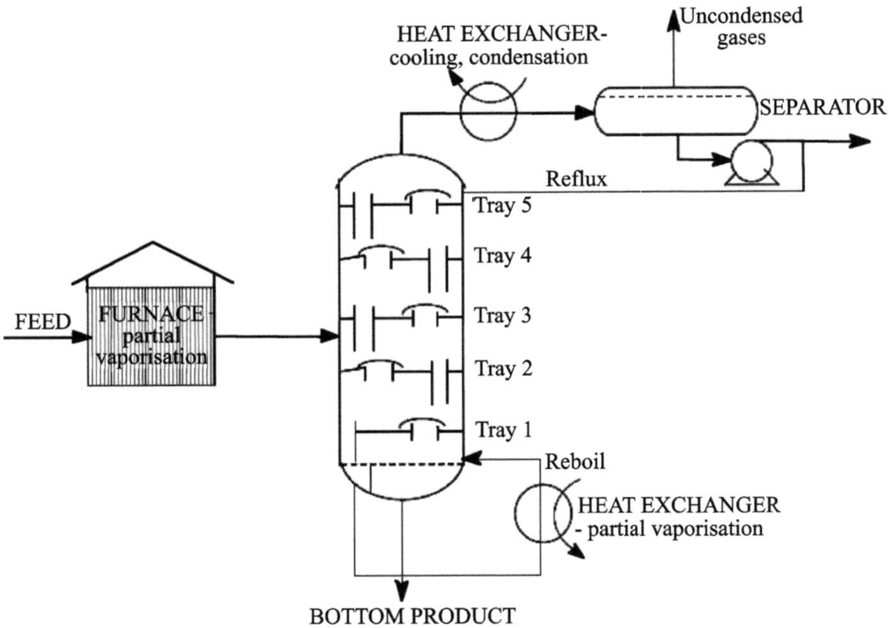

Fig. 3.4 A generalized fractional distillation column

crude oil into various fuels and byproducts, including gasoline, diesel fuel, heating oil, and jet fuel. Main crude oil components are listed in Table 3.4. Since various components boil at different temperatures, refineries use a heating process called distillation to separate the components. For example, gasoline has a lower boiling point than kerosene, allowing the two to be separated by heating to different temperatures. Another important job of the refineries is to remove contaminants from the oil, e. g., removing sulfur from gasoline or diesel to reduce air pollution from the automobile exhausts. After processing at the refinery, gasoline and other liquid products are usually shipped out through pipelines, which are the safest and cheapest way to move large quantities of petroleum across land (Demirbas 2009).

The refining of heavy oil requires exacting and thorough chemical, engineering, and computing processes. Before the actual refining begins, the stored heavy crude oil is cleaned of contaminants such as sand and water.

Industrial distillation is typically performed in large, vertical, steel cylindrical columns known as "distillation towers" or "distillation columns" with diameters ranging from about 65 cm to 11 m and heights ranging from about 6 to 60 m or more. To improve the separation, the tower is normally provided inside with horizontal plates or trays, or the column is packed with a packing material. To provide the heat required for the vaporization, heat is most often added to the bottom of the column by a reboiler.

Large-scale industrial fractionation towers use reflux to achieve more efficient separation of products. Reflux refers to the portion of the condensed overhead liquid product from a distillation tower that is returned to the upper part of the tower. Inside the tower, the downflowing reflux liquid provides cooling and partial con-

Table 3.4 Main crude oil fractions

Component	Boiling range (K)	Number of carbon atoms
Natural gas	<273	C_1 to C_4
Liquefied petroleum gas	231–273	C_3 to C_4
Petroleum ether	293–333	C_5 to C_6
Ligroin (light naphtha)	333–373	C_6 to C_7
Gasoline	313–478	C_5 to C_{12}, and cycloalkanes
Jet fuel	378–538	C_8 to C_{14}, and aromatics
Kerosene	423–588	C_{10} to C_{16}, and aromatics
No. 2 diesel fuel	448–638	C_{10} to C_{22}, and aromatics
Fuel oils	>548	C_{12} to C_{70}, and aromatics
Lubricating oils	>673	$>C_{20}$
Asphalt or petroleum coke	Non-volatile residue	Polycyclic structures

densation of the upflowing vapors, thereby increasing the efficacy of the distillation tower. There are generally 25 to 45 plates or trays in a distillation tower. Each of the plates or trays is at a different temperature and pressure. The stage at the tower bottom has the highest pressure and temperature. Progressing upwards in the tower, the pressure and temperature decreases for each succeeding stage. Another way of improving the separation in a distillation column is to use a packing material instead of trays.

Three major refinery processes change crude oil into finished products: (1) separation, (2) conversion, and (3) purification. The first step is to separate the crude oil into its naturally occurring components. This is known as separation and is accomplished by applying heat through a process called distillation. Separation is performed in a series of distillation towers. The conversion processes have focused on reducing the length of some hydrocarbon chains. The primary purpose of conversion processes is to convert low-value heavy oil into high-value petrol. For example, catalytic reforming is a conversion process. The purpose of the reformer is to increase the octane number of petrol blend components. Once crude oil has been through separation and conversion, the resulting products are ready for purification, which is principally sulfur removal. Common process units found in an oil refinery are presented in Table 3.5.

There is a similar analogy between petroleum refining and biorefining, i.e., by utilizing different processing technologies, components of biomass can be separated and isolated for the production of energy fuels, chemicals, and materials in a biorefinery. Biorefineries can be designed and built to produce desired outputs from the processing of a wide variety of biorenewable materials. These biorefineries will adopt and integrate a range of materials handling and preprocessing equipment, thermochemical and biochemical conversion technologies, and new extraction and purification sciences, to produce a range of intermediate products, while using less energy and reducing effluents and emissions.

Table 3.5 Common process units found in an oil refinery

	Unit	Treatment
1	Desalter	Washes out salt from the crude oil before it enters the atmospheric distillation unit
2	Atmospheric distillation	Distills crude oil into fractions
3	Vacuum distillation	Further distills residual bottoms after atmospheric distillation
4	Naphtha hydrotreater	Uses hydrogen to desulfurize naphtha from atmospheric distillation
5	Catalytic reformer	Used to convert the naphtha-boiling range molecules into higher octane reformer product (reformate)
6	Distillate hydrotreater	Desulfurizes distillates (such as diesel) after atmospheric distillation
7	Fluid catalytic cracker (FCC)	Upgrades heavier fractions into lighter, more valuable products
8	Hydrocracker	Uses hydrogen to upgrade heavier fractions into lighter, more valuable products
9	Visbreaking	Upgrades heavy residual oils by thermally cracking them into lighter, more valuable reduced viscosity products
10	Merox	Treats LPG, kerosene or jet fuel by oxidizing mercaptans to organic disulfides
11	Coking	Converts very heavy residual oils into gasoline and diesel fuel, leaving petroleum coke as a residual product
12	Alkylation	Produces high octane component for gasoline blending
13	Dimerization	Converts olefins into higher octane gasoline blending components
14	Isomerization	Converts linear molecules to higher octane branched molecules for blending into gasoline
15	Steam reforming	Produces hydrogen for the hydrotreaters or hydrocracker
16	Liquefied gas storage	For propane and similar gaseous fuels at pressure sufficient to maintain in liquid form
17	Storage tanks	For crude oil and finished products, usually cylindrical, with some sort of vapor emission control
18	Amine gas and tail gas treatment	For converting hydrogen sulfide from hydrodesulfurization into elemental sulfur
19	Utility units	Such as cooling towers for circulating cooling water, boiler plants for steam generation
20	Wastewater collection and treating	Converts wastewater to water suitable for reuse or for disposal
21	Solvent refining	Uses solvents such as cresol or furfural to remove unwanted, mainly asphaltenic materials from lubricating oil stock
22	Solvent dewaxing	For removing the heavy waxy constituents petroleum from vacuum distillation products

The scale of the biorefining operations will range from medium-sized to very large (equivalent in size to existing chemical plants and pulp and paper mills). Adoption of biorefineries and related processes and product technologies depends on available research, development and prevailing regulations during design and construction.

Biomass, like oil, consists of many different fractions that are separated and made into useful products in biorefineries. However, the processes involved in fractionating biomass are more complex than those used in oil refineries. Another important difference between biorefineries and oil refineries is their size. Biorefineries will most likely be limited in size, because biomass must be produced and transported economically from a limited catchment area. In contrast, oil is drilled and transported all over the world for processing.

Biomass can be processed into plastics, chemicals, fuels, heat, and power in the biorefinery. High-value components, for example essential oils, drugs, or fibers can be recovered as a preprocessing step, with the remaining materials then processed downstream. Processing technologies are most advanced for chemicals and fuels. Biorefineries vary from small single-process plants to large multiple-process sites. Larger biorefineries will be able to integrate different technologies to obtain maximum value from biomass feedstocks. A biorefinery is an integrated plant producing multiple value-added products from a range of renewable feedstocks. This innovative approach responds to changing markets for traditional forest products as well as new products such as energy, chemicals, and materials. The range of feedstocks, processes and potential products is large; each combination of feedstock, process and product is characterized by its own unique combination of technical and economic opportunities, emerging technologies and barriers. Figure 3.5 shows an overview of conversion routes of plant biomass feedstocks to biofuels.

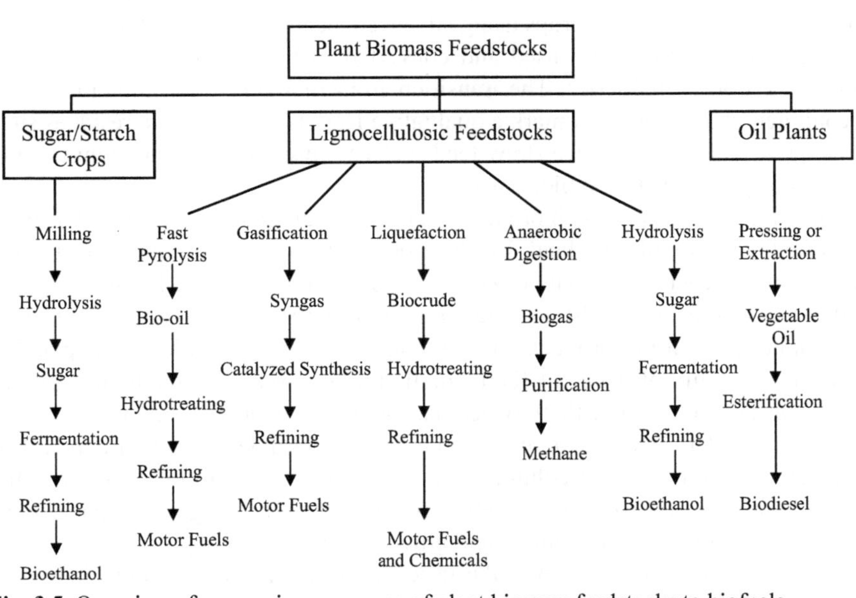

Fig. 3.5 Overview of conversion processes of plant biomass feedstocks to biofuels

Table 3.6 Classification of biorefineries based on their feedstocks

	Feedstocks	Products
Green biorefinery	Grasses and green plants	Ethanol
Cereal biorefinery	Starch crops, sugar crops and grains	Bioethanol
Oilseed biorefinery	Oilseed crops and oil plants	Vegetable oils and biodiesels
Forest biorefinery	Forest harvesting residues, barks, sawdust, pulping liquors and fibers	Fuels, energy, chemicals and materials
Lignocellulosic biorefinery	Agricultural wastes, crop residues, urban wood wastes, industrial organic wastes	Lignocellulosic ethanol, bio-oil and gaseous products

Table 3.6 shows the classification of biorefineries based on their feedstocks. A forest biorefinery will use multiple feedstocks including harvesting residues, extracts from effluents and fractions of pulping liquors to produce fibers, energy, chemicals, and materials. A lignocellulosic-based biorefining strategy may be supported by biomass reserves, created initially with residues from wood product processing or agriculture. Biomass reserves should be used to support first-generation biorefining installations for bioethanol production, development of which will lead to the creation of future high-value co-products (Mabee et al. 2006).

Biorefineries can be classified based on their production technologies as well: first-generation biorefineries (FGBRs), second-generation biorefineries (SGBRs), third-generation biorefineries (TGBRs), and fourth-generation biorefineries.

The FGBRs refer to biofuels made from sugar, starch, vegetable oils, or animal fats using conventional technology. Second and third-generation biorefineries are also called advanced biorefineries. SGBRs made from non-food crops, wheat straw, corn, and wood are energy crops using advanced technology (see Table 3.7).

Sugar and vegetable are used and converted to bioalcohols and biodiesel in first-generation biorefineries. The transition from first-generation biorefineries to second-generation ones will mark a qualitative leap. Lignocellulosic residues such as sugarcane bagasse and rice straw feedstocks are used and they are converted to SGBs in second-generation biorefineries.

The first third-generation biorefinery demonstration plant in the world has been commissioned in Oulu, Finland, by Chempolis Oy, and will produce paper fiber, biofuel and biochemicals from non-wood and non-food materials. TGBRs will start with a mix of biomass feedstocks (agricultural or forest biomass) and will produce a multiplicity of various products, such as ethanol for fuels, chemicals, and plastics, by applying a mix of different (both small and large-scale) technologies such as extraction and separation, thermochemical or biochemical conversion. However, large integrated TGBRs are not expected to become established until around 2020. Increasing quantities of agricultural residues will be needed to make paper in the future, as insufficient wood is available locally in the world's growing paper markets, forest resources are declining, and growing environmental pressures are being put on the use of wood. Vegetable oil is used and converted to biogasoline in fourth-generation biorefineries.

Table 3.7 Classification of biorefineries based on their generation technologies

Generation	Feedstocks	Examples
First-generation biorefineries	Sugar, starch, vegetable oils, or animal fats	Bioalcohols, vegetable oil, biodiesel, biosyngas, biogas
Second-generation biorefineries	Non-food crops, wheat straw, corn, wood, solid waste, energy crop	Bioalcohols, bio-oil, bio-DMF, biohydrogen, bio-Fischer–Tropsch diesel
Third-generation biorefineries	Algae	Vegetable oil, biodiesel
Fourth-generation biorefineries	Vegetable oil, biodiesel	Biogasoline

Table 3.8 Classification of biorefineries based on their conversion routes

Biorefinery	Products
Biosyngas-based	Syngas, hydrogen, methanol, dimethyl ether, FT diesel
Pyrolysis-based	Bio-oil, diesel fuel, chemicals, oxygenates, hydrogen
Hydrothermal upgrading-based	C_xH_x, diesel fuel, chemicals
Fermentation-based	Bioethanol
Oil plant-based	Biodiesel, diesel fuel, gasoline

Biorefineries can be classified based on their conversion routes: biosyngas-based biorefineries, pyrolysis-based biorefineries, hydrothermal upgrading-based biorefineries, fermentation-based biorefineries, and oil plant-based biorefineries. Table 3.8 shows the classification of biorefineries based on their conversion routes. Biosyngas is a multi-functional intermediate for the production of materials, chemicals, transportation fuels, power and/or heat from biomass.

References

Balat, M. 2007. Production of biodiesel from vegetable oils: A survey. Energy Sources Part A 29:895–913.

Balat, M. 2008. Progress in biogas production processes. Energy Edu Sci Technol 22:15–35.

Balat, M. 2009. New biofuel production technologies. Energy Edu Sci Technol Part A 22:147–161.

Chhetri, A.B., Islam, M.R. 2008. Towards producing a truly green biodiesel. Energy Sources Part A 30:754–764.

Demirbas, A. 2002. Biodiesel from vegetable oils via transesterification in supercritical methanol. Energy Convers Manage 43:2349–2356.

Demirbas, A. 2003. Biodiesel fuels from vegetable oils via catalytic and non-catalytic supercritical alcohol transesterifications and other methods: a survey. Energy Convers Manage 44:2093–2109.

Demirbas, A. 2007. Converting biomass derived synthetic gas to fuels via Fischer–Tropsch synthesis. Energy Sources Part A 29:507–512.

Demirbas, A. 2008. Recent progress in biorenewable feedstocks. Energy Edu Sci Technol 22:69–95.

Demirbas, A. 2009. Biofuels: securing the planet's future energy needs. Springer, London.

Demirbas, A., Karslioglu, S. 2007. Biodiesel production facilities from vegetable oils and animal fats. Energy Sources Part A 29:133–141.

Dincer, K. 2008. Lower emissions from biodiesel combustion. Energy Sources Part A 30:963–968.

GBEP. 2007. Review of current state of biorefinery development in G8+5. Global Biorefinery Partnership, New York.

Goldemberg, J. 2000. World Energy Assessment. United Nations Development Programme, New York.

Kamm, B., Kamm, M., Gruber, P. (Eds.) 2005. Biorefineries: biobased industrial processes and products. Status quo and future directions. Wiley-VCH, Weinheim.

Keskin, A. 2009. Biodiesel production from free fatty acids obtained with neutralization of the crude glycerin. Energy Sources Part A 31:17–24.

Khoiyangbam, R.S. 2008. Greenhouse gases emission from energy production in conventional biogas plants in India. Energy Sources Part A 30:689–697.

Mabee, W.E., Fraser, E.D.G., McFarlane, P.N., Saddler, J.N. 2006. Canadian biomass reserves for biorefining. Appl Biochem Biotechnol 129:22–40.

Parikka, M. 2004. Global biomass fuel resources. Biomass Bioenergy 27:613–620.

WI (Worldwatch Institute). 2007. Biofuels for transport: global potential and implications for energy and agriculture. Earthscan, London.

Chapter 4
Transportation Fuels

4.1 Introduction

Today, most of the energy we use comes from fossil fuels: petroleum, coal, and natural gas. Fossil fuels are used mainly in electric and transportation industries. All fossil fuels release greenhouse gases and other pollutants into the atmosphere. There is a growing concern that the world may run out of petroleum-based fuel resources. All these make it imperative that the search for alternative fuels is made in earnest. The diesel engine is used mainly for heavy vehicles. The diesel process means that air is compressed in the engine and then the fuel is injected and ignited by the hot, compressed air.

Ethanol from biomass and Fischer–Tropsch liquids from natural gas may become widespread fuels. The Fischer–Tropsch liquids will penetrate the market if there are large amounts of stranded natural gas selling for very low prices at the same time that petroleum is expensive or extremely low sulfur is required in diesel fuel. Alternative fuels for Otto engines or light-duty vehicles (LDVs, e. g., cars and light trucks) contain (1) reformulated gasoline, (2) compressed natural gas, (3) methanol and ethanol, (4) liquid petroleum gas, (5) liquefied natural gas, (6) Fischer–Tropsch liquids using syngas from natural gas, (7) hydrogen, (8) electricity, and (f) fuel cells (MacLean and Lave 2003). Alternative fuels for diesel engines or heavy-duty vehicles (HDEs) contain Fischer–Tropsch (FT) diesel, vegetable oil, biodiesel, and dimethyl ether (DME). Both FT and DME can be manufactured from natural gas and are therefore not limited by feedstock availability. Biodiesel on the other hand, is produced from vegetable oils whose supply for non-nutritional uses is presently quite limited.

4.2 Gasoline or Petrol

Gasoline, a petroleum-derived liquid mixture, is primarily used as fuel in internal combustion engines, specifically in spark-ignition engines. This type of engine, also known as Otto cycle engine, uses a compressed mixture of gasoline and air, which is then ignited by a spark plug.

Gasoline is a refined product of petroleum consisting of a mixture of over 500 hydrocarbons with 5 to 12 carbons, as well as additives, and blending agents. The composition of gasolines varies widely, depending on the crude oils used, the refinery processes available, the overall balance of product demand, and the product specifications. For example, the blend of hydrocarbons may include some contaminants, such as sulfur, nitrogen, oxygen, and certain metals. Gasoline however consists mostly of aliphatic hydrocarbons, enhanced with iso-octane or the aromatic hydrocarbons toluene and benzene to increase its octane rating. The four major constituent groups of gasoline are olefins, aromatics, paraffins, and napthenes.

The amount of gasoline that can be distilled from crude depends on where the crude is obtained. Venezuelan crude yields little gasoline (about 5%), whereas Texas or Arabian crude yields about 30% gasoline, which is called "straight run" gasoline. Average gasoline yields at US refineries averaged to 45.5% in 2007.

The important characteristics of gasoline are density, vapor pressure, distillation range, octane number, and chemical composition, as explained below. To be attractive, a motor gasoline must have (a) desirable volatility, (b) anti-knock resistance (related to octane rating), (c) good fuel economy, (d) minimal deposition on engine component surfaces, and (e) complete combustion and low pollutant emissions (Chigier 1981).

The density of gasoline is 0.71–0.77 kg/L. Gasoline is more volatile than diesel oil, Jet-A or kerosene, not only because of the base constituents, but also because of the additives that are put into it. The final control of volatility is often achieved by blending with butane. The desired volatility depends on the ambient temperature: in hotter climates, gasoline components of higher molecular weight and thus lower volatility are used. In cold climates, too little volatility results in cars failing to start. In hot climates, excessive volatility results in what is known as "vapor lock" where combustion fails to occur, because the liquid fuel has changed to a gaseous fuel in the fuel lines, rendering the fuel pump ineffective and starving the engine of fuel.

Another important characteristic of gasoline is its octane number or octane rating, which is a measure of how resistant gasoline is to the abnormal combustion phenomenon known as predetonation (also known as knocking, pinging, spark knock, and other names). Octane number is measured relative to a mixture of 2,2,4-trimethylpentane and *n*-heptane. Higher octane numbers are preferred in internal combustion engines.

Table 4.1 shows the typical composition of gasoline. The typical composition of gasoline hydrocarbons (% volume) is as follows: 4–8% alkanes, 2–5% alkenes, 25–40% iso-alkanes, 3–7% cycloalkanes, 1–4% cycloalkenes, and 20–50% total aromatics (0.5–2.5% benzene). Additives and blending agents are added to the hydrocarbon mixture to improve the performance and stability of gasoline. These

Table 4.1 Typical composition of gasoline

General name	Examples	Composition (wt.%)
Aliphatic: straight chain	Heptane	15–25
Aliphatic: branched	Iso-octane	15–25
Aliphatic: cyclic	Cyclopentane	20–30
Aromatic	Ethyl benzene	20–30

compounds include anti-knock agents, anti-oxidants, metal deactivators, lead scavengers, anti-rust agents, anti-icing agents, upper-cylinder lubricants, detergents, and dyes. Physical and chemical properties of gasoline are given in Table 4.2. Table 4.3 shows the major components of gasoline.

Only compressed natural gas (CNG) and liquid petroleum gas (LPG) appear to have some economic advantage relative to gasoline while ethanol, methanol, hydrogen and electricity are at a severe economic disadvantage (Piel 2001).

Table 4.2 Physical and chemical properties of gasoline

Property	Information
Color	Colorless to pale brown or pink
Average molecular weight	108
Density (kg/L)	0.7–0.8
Flash point (K)	227.2
Explosive limits in air	1.3–6.0%
Flammability limits	1.4–7.4%
Auto-ignition (K)	553–759
Boiling point (K)	
Initially	312
After 10% distillate	333
After 50% distillate	383
After 90% distillate	443
Final boiling point	477
Solubility	
Water at 293 K	Insoluble
Absolute ethanol	Soluble
Diethyl ether	Soluble
Chloroform	Soluble
Benzene	Soluble

Table 4.3 Major components of gasoline

Component	Composition (wt.%)
n-Alkanes	
C_5	3.0
C_6	11.6
C_7	1.2
C_9	0.7
C_{10}–C_{13}	0.8
Total of *n*-alkanes	17.3
Branched alkanes	
C_4	2.2
C_5	15.1
C_6	8.0
C_7	1.9
C_8	1.8
C_9	2.1
C_{10}–C_{13}	1.0
Total of branched alkanes	32.0
Cycloalkanes	
C_6	3.0
C_7	1.4
C_8	0.6
Total of cycloalkanes	5.0
Olefins	
C_6	1.8
Total of olefins	1.8
Aromatics	
Benzene	3.2
Toluene	4.8
Xylenes	6.6
Ethylbenzene	1.4
C_3 benzenes	4.2
C_4 benzenes	7.6
Others	2.7
Total aromatics	30.5

Ethanol has a higher octane number (108), broader flammability limits, higher flame speeds and higher heats of vaporization than gasoline. These properties allow for a higher compression ratio, shorter burn time and leaner burn engine, which lead to theoretical efficiency advantages over gasoline in an ICE. Disadvantages of ethanol include its lower energy density than gasoline, its corrosiveness, low flame luminosity, lower vapor pressure, miscibility with water, and toxicity to ecosystems.

Currently, ethanol is produced from sugar beets and from molasses. A typical yield is 72.5 L of ethanol per ton of sugar cane. Modern crops yield 60 tons of sugar cane per hectare of land. Production of ethanol from biomass is one way to reduce both the consumption of crude oil and environmental pollution (Lang et al. 2001). Domestic production and use of ethanol for fuel can decrease dependence on foreign oil, reduce trade deficits, create jobs in rural areas, reduce air pollution, and reduce global climate change with less carbon dioxide buildup (Demirbas 2009).

Methanol can be used as one possible replacement for conventional motor fuels. Methanol has been seen as a possible large-volume motor fuel substitute at various times during gasoline shortages. It was often used in the early part of the century to power automobiles before inexpensive gasoline was widely introduced. Methanol is poisonous and burns with an invisible flame. Methanol has, just like ethyl alcohol, a high octane rating and hence an Otto engine is preferable. If an ignition booster is used, methanol can be used in a diesel engine.

Natural gas (NG) consists of 85–95% of methane, CH_4, which is the simplest hydrocarbon. NG is the cleanest burning alternative fuel. Exhaust emissions from NG vehicles are much lower than those from gasoline-powered vehicles. Combustion of NG reduces carbon dioxide content compared with diesel, but the lost efficiency when the Otto process is used means that carbon dioxide emissions increase. All in all, combustion of NG in a gasoline engine gives rise to about as much carbon dioxide as the combustion of diesel in a diesel engine, measured in units of energy.

Even though NG is a cleaner burning fuel than either gasoline or diesel, it also contains active compounds, such as sulfur, and inert compounds such as nitrogen and CO_2. NG has a high octane number (110–130) and therefore potential for use in a high-compression engine. NG is a fossil fuel and is extracted from gas sources in bedrock. It is lighter than air and as such carries a certain risk of explosion.

P-series fuel is a unique blend of natural gas liquids, ethanol, hydrocarbons methyltetrahydrofuran (MeTHF). MeTHF is a biomass-derived co-solvent. P-series is made primarily from renewable resources and provides significant emissions benefits over reformulated gasoline. P-series fuels are clear, colorless, liquid blends with octane numbers of 89–93, which are formulated to be used in flexible fuel vehicles. Like gasoline, low vapor pressure formulations are produced to prevent excessive evaporation during summer and high vapor pressure formulations are used for easy starting in the winter. P-series fuels are at least 60% non-petroleum. Because a majority of the components that make up P-series fuels come from domestically produced renewable resources, this alternative fuel promotes both energy security and environmental quality. Theoretically, these fuels could be 96% derived from domestic resources. P-series fuels could reduce fossil energy use by 49 to 57% and petroleum use by 80% relative to gasoline. Greenhouse gas emissions of the P-series fuels are 45 to 50% below those of reformulated gasoline. P-series would be used in severe cold-weather conditions to meet cold start requirements.

P-series also have many environmental benefits. Emissions from the production and use of P-series are substantially better than those from gasoline. Each unit of

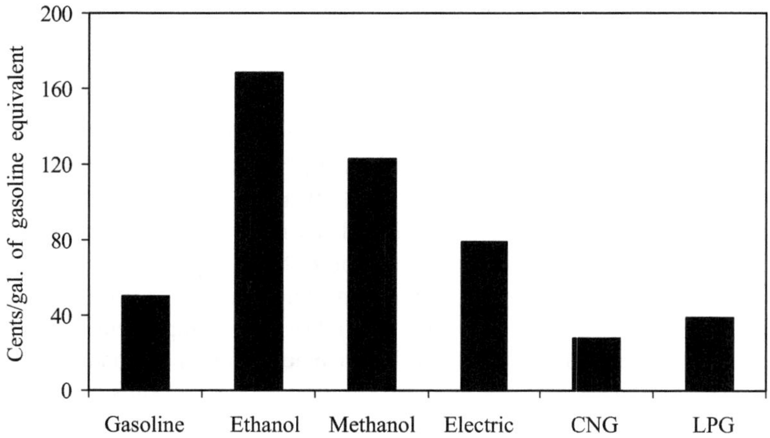

Fig. 4.1 Economics of gasoline alternative fuels

P-series fuel emits approximately 50% less carbon dioxide, 35% less hydrocarbons, and 15% less carbon monoxide than gasoline. It also has 40% less ozone-forming potential.

Electricity can be used as a transportation fuel to power battery electric and fuel cell vehicles. Electric fuel is electricity that is used to directly power the vehicle. The power is created by batteries and other electricity sources. Batteries are energy storage devices, but unlike batteries, fuel cells convert chemical energy to electricity. Fuel cell vehicles use electricity produced from an electrochemical reaction that takes place when hydrogen and oxygen are combined in the fuel cell stack. The production of electricity using fuel cells takes place without combustion or pollution and leaves only two byproducts: heat and water. Benefits include no emissions and fewer parts to be serviced and replaced. Electricity is also cheaper than gasoline.

Solar energy technologies use sunlight to produce heat and electricity. Solar fuel is light from the Sun transformed into energy. Electricity produced by solar energy through photovoltaic technologies can be used in conventional electric vehicles. Solar energy cannot be used directly in cars except as a publicity stunt. Solar power may be used to run certain auxiliary systems in the vehicle. In order to collect this energy and use it to fuel a vehicle, photovoltaic cells are used. Pure solar energy is 100% renewable and a vehicle run on this fuel emits no emissions. Figure 4.1 shows the wholesale prices of a number of possible alternative fuels on an energy equivalent basis compared to conventional gasoline (AICHE 1997).

The most commonly used measure of a gasoline's ability to burn without knocking is its octane number. Table 4.4 shows hydrocarbon octane numbers. By 1922 a number of compounds had been discovered that increased the octane number of gasoline. Adding as little as 6 mL of tetraethyllead (Fig. 4.2) to a gallon of gasoline, for example, can increase the octane number by 15 to 20 units. This discovery gave rise to the first "ethyl" gasoline, and enabled the petroleum industry to produce aviation gasolines with octane numbers greater than 100.

About 10% of the product of the distillation of crude oil is a fraction known as straight-run gasoline. Straight-run gasoline burns unevenly in high-compression

Table 4.4 Hydrocarbon octane numbers

Hydrocarbon	Octane number (ON)
Heptane	0
2-Methylheptane	23
Hexane	25
2-Methylhexane	44
1-Heptene	60
Pentane	62
1-Pentene	84
Butane	91
Cyclohexane	97
2,2,4-Trimethylpentane (iso-octane)	100
Benzene	101
Toluene	112

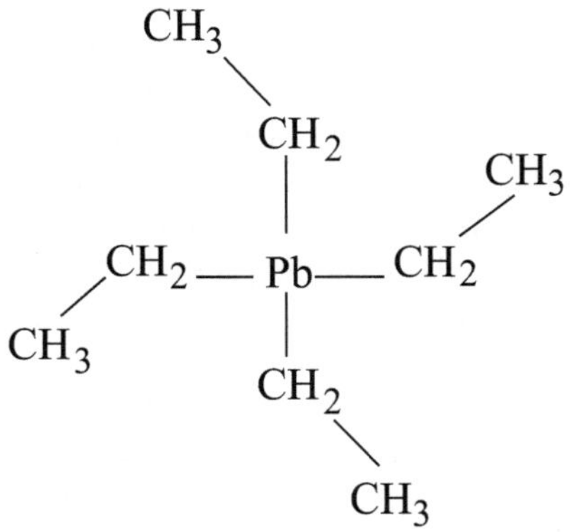

Fig. 4.2 Chemical structure of tetraethyllead

engines, producing a shock wave that causes the engine to "knock" or "ping." The relationship between knocking and the structure of the hydrocarbons in gasoline is summarized in the following general rules:

- Branched alkanes and cycloalkanes burn more evenly than straight-chain alkanes.
- Short alkanes (C_4H_{10}) burn more evenly than long alkanes (C_7H_{16}).
- Alkenes burn more evenly than alkanes.
- Aromatic hydrocarbons burn more evenly than cycloalkanes.

4.3 Diesel Fuel

The crude oil is separated by boiling points into six main grades of hydrocarbons: refinery gas (used for refinery fuel), gasoline (naphthas), kerosene, light oils (diesel oil or diesel fuel) and heavy gas oils (fuel oil), and long residue. This initial separation is done by distillation. The first step in the refining of crude oil, whether in a simple or a complex refinery, is the separation of the crude oil into fractions, known as fractionation or distillation. These fractions are mixtures containing hydrocarbon compounds whose boiling points lie within a specified range.

Almost all motor vehicles today are powered by either gasoline or diesel. Both fuels are derived from petroleum, a non-renewable resource. Diesel fuel is produced by distilling raw oil, which is extracted from bedrock. Diesel fuel consists of hydrocarbons with between 9 and 27 carbon atoms in a chain as well as a smaller amount of sulfur, nitrogen, oxygen and metal compounds. It is a general property of hydrocarbons that the auto-ignition temperature is higher for more volatile hydrocarbons. Alkanes, naphthenes, olefins, and aromatics are all hydrocarbons present in diesel fuels. In addition, other substances are added to improve the characteristics of diesel fuel. A good diesel fuel is characterized by the following characteristics: low sulfur and aromatic content, good ignition quality, the right cold-weather properties, low content of pollutants, as well as the right density, viscosity and boiling point.

There are several different grades of diesel, depending upon its intended use. Like gasoline, diesel fuel is not a single substance, but a mixture of various petroleum-derived components, each with their own physical and chemical properties.

Cetane number, volatility, and viscosity are the most important characteristics of diesel fuel. Cetane number (CN) is a measure of the ignition quality of the fuel, and is based on two compounds, namely hexadecane with a CN of 100 and heptamethylnonane with a CN of 15. CN affects combustion roughness. Consumers often think the CN is similar to the octane number for gasoline, but that is not the case. Octane is a measure of a spark-ignition engine fuel's (gasoline) ability to resist engine knock (preignition from compression). Diesel cetane ratings work in the opposite direction. The higher the cetane rating, the more easily it ignites. Reaching desired cetane levels also limits the aromatic content of diesel fuel. Diesel fuel cetane ratings are calculated by calibrating a fuel to a mixture of reference fuels in a specially designed cooperative fuel research (CFR) engine. Various tests have been developed to calculate the CN from certain fuel properties. These tests usually involve some combination of fuel density and distillation properties. The two more common CN estimate formulas, referred to as cetane indices, are used to distinguish their results from the engine test, known as the ASTM D 976 and ASTM D 4737.

Unlike spark-ignition engines, the power and economy of diesel engines are comparatively insensitive to fuel volatility. There is some indirect impact in that less-volatile fuels have higher heating values (HHV). Conversely fuels with higher front-end volatility tend to improve starting and warm-up performance and reduce smoke. Ideal fuel volatility requirements will vary based on engine size and design, speed and load conditions, and atmospheric conditions. As an example, more volatile fuels may provide better performance for fluctuating loads and speeds such as those experienced by trucks and buses.

The viscosity of diesel fuel is an important property, which impacts the performance of fuel injection systems. Some injection pumps can experience excessive wear and power loss due to injector or pump leakage if viscosity is too low. If fuel viscosity is too high, it may cause too much pump resistance, filter damage and adversely affect fuel spray patterns. High fuel viscosity can cause an injector spray pattern with poor fuel dispersion.

Using syngas made from coal, researchers were able to produce liquid hydrocarbons rich in oxygenated compounds in what was termed the synthol process. Succeeding these initial discoveries, considerable effort went into developing improved and more selective catalysts for this process.

The first Fischer–Tropsch (FTS) plants began operation in Germany in 1938 but closed down after World War II. Then in 1955, Sasol, a world-leader in the commercial production of liquid fuels and chemicals from coal and crude oil, started Sasol I in Sasolburg, South Africa. Following the success of Sasol I, Sasol II and III, located in Secunda, South Africa, came on line in 1980 and 1982, respectively (Spath and Mann 2000; Spath and Dayton 2003). The FTS is an essential step to convert carbon-containing feedstocks to liquid fuels such as diesel. Major advantages of the FTS are: (1) flexibility in feedstocks (natural gas, coal, biomass), (2) large and even sustainable resources, (3) ultraclean (low sulfur content) products, and (4) suitable for converting difficult-to-process resources. A major drawback of the FT process is the polymerization-like nature of the process, yielding a wide product spectrum, ranging from compounds with low molecular mass such as methane, to products with very high molecular mass like heavy waxes. The industrial application of the FT process started in Germany and by 1938 there were nine plants in operation having a combined capacity of about 660×10^3 ton per year (Anderson 1984).

The high-temperature fluidized bed FT reactors with iron catalyst are ideal for the production of large amounts of linear olefins. As petrochemicals, they sell at much higher prices than fuels. The olefin content of the C_3, C_5–C_{12} and C_{13}–C_{18} cuts are typically 85, 70 and 60%, respectively (Dry 2002).

The FTS-based gas to liquids (GTL) technology includes the three processing steps, namely syngas generation, syngas conversion, and hydroprocessing. In order to make the GTL technology more cost-effective, the focus must be on reducing both the capital and the operating costs of such a plant (Vosloo 2001). For some time now the price has been up to $60 per barrel. It has been estimated that the FT process should be viable at crude oil prices of about $20 per barrel (Jager 1998; Dry 2004). The current commercial applications of the FT process are geared at the production of the valuable linear alpha olefins and of fuels such as LPG, gasoline, kerosene and diesel. Since the FT process produces predominantly linear hydrocarbons the production of high-quality diesel fuel is currently of considerable interest (Dry 2004). The most expensive section of an FT complex is the production of purified syngas, and so its composition should match the overall usage ratio of the FT reactions, which in turn depends on the product selectivity (Dry 2002).

Recently, there has been some interest in the use of FTS for biomass conversion to synthetic hydrocarbons. Biomass can be converted to biosyngas by non-catalytic, catalytic and steam gasification processes.

Gasification followed by Fischer–Tropsch synthesis (FTS) is currently the most promising method for upgrading low-value coal and biomass to high-value liquid fuels and chemicals. The total biomass produced each year as waste material from agriculture and forest operations could thus be converted into roughly 40 billion gallon per year of liquid fuels, roughly 25% of the current US gasoline usage.

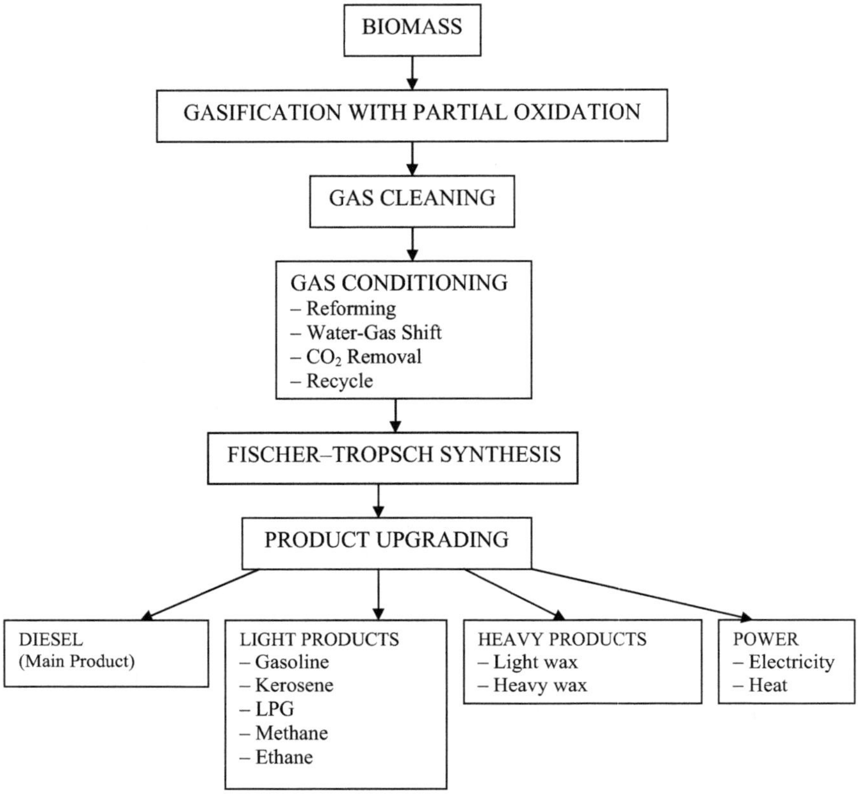

Fig. 4.3 Diesel and other products from biomass via the Fischer–Tropsch synthesis

Tijmensen et al. (2002) review the technical feasibility and economics of the biomass-integrated gasification Fischer–Tropsch (BIG-FT) process and also point out the key R&D issues involved in commercialization of this process. Boerrigter and den Uil (2002) give a similar review identifying a potential BIG-FT process configuration. The FTS for the production of liquid hydrocarbons from coal-based synthesis gas has been the subject of renewed interest for conversion of coal and natural gas to liquid fuels (Jin and Datye 2000).

To produce biosyngas from a biomass fuel the following procedures are necessary: (a) gasification of the biomass, (b) cleaning of the product gas, and (c) usage of the synthesis gas to produce liquid fuel via the Fischer–Tropsch synthesis. Figure 4.3 shows diesel and other products from biomass via the Fischer–Tropsch synthesis.

4.4 Liquefied Petroleum Gas

Liquefied petroleum gas (LPG) mainly contains propane (C_3H_8) and butane (C_4H_{10}). Both propane and butane are easily liquefied and stored in pressure containers. These properties make the fuel highly portable, and hence, it can be easily transported in cylinders or tanks to end users. LPG is a gas blend of gases produced commercially in petroleum refineries and stored under pressure below critical temperature in order to keep it in a liquid state. LPG also can be produced from natural gas. The normal boiling point of LPG varies from 229 to 273 K so that the pressure required liquefying. LPG is an attractive fuel for internal combustion engines because it burns with little air pollution and little solid residue, and also because it does not dilute lubricants, and has a high octane rating.

Unlike gasoline, LPG is used as a dry gas and contains no fuel additives. Without lead as a lubricating agent, some early LPG conversions resulted in increased wear rates in the engine cylinder valves. LPG has higher ignition temperatures than gasoline, which increases the importance of maintaining proper ignition system operation, and reduces improper combustion and sluggish vehicle performance.

The LPG fuel/air mixture burns well, so that engine starting problems are less common than with liquid fuels. LPG also has a lower energy density but higher octane rating than gasoline. Consequently, LPG produces fewer miles per gallon but the octane value allows higher compression ratios and higher power and fuel efficiency. For the same compression ratio, the amount of gaseous LPG air/fuel mixture that can be burned is about 5–10% less than in gasoline engines.

LPG is a good substitute for petrol in spark-ignition engines. Its clean burning properties, in a properly tuned engine, give reduced exhaust emissions, as well as extended lubricant and spark plug life. LPG is one of the cleanest alternative fuels for motor vehicles. LPG presents a useful combination of combustion and storage properties that makes it an attractive vehicular fuel.

4.5 Compressed Natural Gas

Natural gas can be stored in compressed form and is then known as compressed natural gas (CNG), or at a temperature (111 K) in liquid form, known as liquefied natural gas (LNG). LNG has been considered as an option for fueling automobiles.

Despite the large difference in volumetric energy density between gasoline and CNG, the impact of CNG energy density on engine performance is less dramatic. As a gas, it has few cold-start problems. Its higher octane value allows for higher engine compression ratios than can be used with gasoline alone. Higher compression ratios allow for higher power and fuel efficiency. However, for the same compression ratio, the amount of natural gas air/fuel mixture that can be burned in each piston stroke is 10–15% less than for gasoline. Thus, there is a 10–15% loss of engine output power.

To use CNG safely, technicians and drivers need to know what the differences are and how to work with them. Other subjects that must be understood are NG combustion and storage, working with high-pressure conduits, connectors, regulators, and cylinders, safety codes and industry standards, and recommended CNG cylinder inspections. Detailed hands-on training for installation and maintenance technicians should normally be provided by the conversion kit manufacturers.

The toxic emissions with CNG, without exception, are lower than for any other hydrocarbon fuel. This is a direct result of the fact that CNG is a single hydrocarbon, 90% methane, whereas all of the other fuels are a mix of hydrocarbons. LPG is a relatively simple mix of propane, butane, and pentane compared to CNG and the complex mix that makes up the gasoline and diesel typically purchased at the service station. Gasoline and diesel emit compounds into the air: methanol, formaldehyde, aldehydes, acrolein, benzene, toluene, xylene, etc., some of which are not yet part of any established emission standard but certainly are not beneficial to people's health.

4.6 Hydrogen

The combustion products of hydrogen when it is burned completely with air consist of water, oxygen, and nitrogen. In the meantime, it has been suggested, hydrogen is too valuable to burn. Laboratory tests conducted on internal combustion engines burning hydrogen demonstrate good performance (Berry et al. 1996). In comparison with an engine burning gasoline, the emission of nitrogen oxides is far less for the engine-fueled hydrogen. The product of hydrogen combustion with air is water vapor and negligible pollution when the peak temperature is limited. Some oxides of nitrogen (NO_x) are formed at very high combustion temperatures ($<2300\,K$); fortunately, the auto-ignition temperature of hydrogen is only 858 K.

Hydrogen has good properties as a fuel for internal combustion engines in automobiles. Some of the characteristic properties of a hydrogen-air mixture that can definitely influence the engine design and performance are low ignition energy, low density, wide range of ignition limits, high diffusion speed, and high flame speed (Plass et al. 1990).

The main disadvantages of using hydrogen as a fuel for automobiles are huge on-board storage tanks, which are required because of hydrogen's extremely low density. Hydrogen may be stored on board a vehicle as compressed gas in ultrahigh-pressure vessels, as a liquid in cryogenic containers, or as a gas bound with certain metals in metal hydrides.

Hydrogen is one of the most promising alternative energy technologies. Hydrogen can be generated in a number of ways, such as electrochemical processes, thermochemical processes, photochemical processes, photocatalytic processes, or photoelectrochemical processes (Momirlan and Veziroglu 1999; Momirlan and Veziroglu 2002). Biohydrogen production by anaerobic fermentation from renewable organic waste sources has been found to be a promising method for the recovery of bioenergy (Han and Shin 2004). In this method, anaerobic bacteria use organic substances as the sole source of electrons and energy, converting them into hydrogen.

The use of hydrogen as a fuel for transportation and stationary applications is receiving much favorable attention as a technical and policy issue (Cherry 2004). Hydrogen gas is being explored for use in combustion engines and fuel-cell electric vehicles. It is a gas at normal temperatures and pressures, which presents greater transportation and storage hurdles than exist for the liquid fuels. Several hydrogen technologies are under development; the most promising of these is the fuel cell. Fuel cells use hydrogen, oxygen, catalysts, and electrolytic solution to produce energy in the form of heat and electricity.

4.7 Biorenewable Liquid Fuels

The most important biorenewable liquid fuels are bioethanol and biodiesel. Bioethanol is a petrol additive/substitute. Biodiesel is a diesel alternative. Biorenewable fuels are safely and easily biodegradable and so are particularly well suited to be environmentally friendly. Biodiesel, a biofuel that can replace directly petroleum-derived diesel without engine modifications, has gained a lot of attention due to its environmental and technological advantages.

Production of motor fuel alternatives from biomass materials is an important application area of biotechnological methods. Table 4.5 shows the potential and available motor fuels. Biorenewable sourced motor fuel alternatives are:

1. Gasoline–alcohol mixtures
2. Alcohol substituting gasoline
3. Gasoline–vegetable oil mixtures
4. Diesel fuel–vegetable oil mixtures
5. Vegetable oil substituting diesel fuel

In gasoline–alcohol mixtures ethanol and methanol are generally used, and in gasoline engines mixtures containing 20% or less alcohol by volume can be used without altering the construction of the engine. Because of the hygroscopic properties of ethanol and methanol the gasoline–alcohol mixtures are in fact ternary mixtures composed of gasoline–alcohol and water. In the evaluation of such mixtures as motor fuel there is the phase separation problem, which depends on several factors. It is seen in the literature that there are numerous attempts to overcome this problem (Mislavskaya et al. 1982; Osten and Sell 1983).

In gasoline–methanol mixtures containing 0.1% water, *i*-propanol is added to the environment (medium) in order to decrease the phase separation temperature. Fuels containing different ratios of gasoline–methanol–*i*-propanol and water are also

Table 4.5 Potential and available motor fuels

Fuel type	Available motor fuel
Traditional fuels	Diesel and gasoline
Oxygenated fuels	Ethanol 10% (E10), methanol, methyl tertiary butyl ether (MTBE), ethyl tertiary butyl ether (ETBE), tertiary butyl alcohol (TBA), and tertiary amyl methyl ether (TAME)
Alternative fuels	Liquefied petroleum gases (LPG), ethanol 85% (E85), ethanol, 95% (E95), methanol 85% (M85), methanol, neat (M100), compressed natural gas (CNG), liquefied natural gas (LNG), biodiesel (BD), hydrogen, and electricity

composed, which are proven to be stable in the climatic conditions. An increase in the aromatic character of the gasoline, a decrease in the water content of the mixture, and an increase in the amount of the additive used results in a decrease in the phase separation temperature of the mixture. In gasoline–ethanol mixtures the additive used is also *i*-propanol. In gasoline–alcohol mixtures various additives are used like *i*-propanol, *n*-butanol, *i*-butanol, and *i*-amylalcohol.

Ethanol is a liquid biofuel, which can be produced from several different biomass feedstocks and conversion technologies. Bioethanol is an attractive alternative fuel because it is a renewable bio-based resource and it is oxygenated, thereby providing the potential to reduce particulate emissions in compression ignition engines.

Ethanol can be made synthetically from petroleum or by microbial conversion of biomass materials through fermentation. In 1995, about 93% of the ethanol in the world was produced by the fermentation method and about 7% by the synthetic method. The fermentation method generally uses three steps: (1) the formation of a solution of fermentable sugars, (2) the fermentation of these sugars to ethanol, and (3) the separation and purification of the ethanol, usually by distillation.

Ethanol has a higher octane number, broader flammability limits, higher flame speeds and higher heats of vaporization than gasoline. These properties allow for a higher compression ratio, shorter burn time and leaner burn engine, which lead to theoretical efficiency advantages over gasoline in an internal combustion engine. Disadvantages of ethanol include its lower energy density than gasoline (but about 35% higher than that of methanol), its corrosiveness, low flame luminosity, lower vapor pressure (making cold starts difficult), miscibility with water, and toxicity to ecosystems.

Because ethanol contains oxygen (35% of oxygen content) in the fuel, it can effectively reduce particulate matter emission in the diesel engine. Ethanol is appropriate for the mixed fuel in the gasoline engine because of its high octane number, and its low cetane number, and the high heat of vaporization impedes self-ignition in the diesel engine. The most popular blend for light-duty vehicles is known as E85, and contains 85% ethanol and 15% gasoline. In Brazil, bioethanol for fuel is derived from sugarcane and is used pure or blended with gasoline in a mixture called gasohol (24% bioethanol, 76% gasoline). In several states of the US, a small amount of bioethanol (10% by volume) is added to gasoline, known as gasohol or E10. Blends having higher concentrations of bioethanol in gasoline are also used, e.g., in flexible fuel vehicles that can operate on blends of up to 85% bioethanol (E85). Some countries have exercised biofuel programs for bioethanol–gasoline blends such as the US (E10 and for flexible fuel vehicles (FFV) E85), Canada (E10 and for FFV E85), Sweden (E5 and for FFV E85), India (E5), Australia (E10), Thailand (E10), China (E10), Columbia (E10), Peru (E10), Paraguay (E7), and Brazil (E20, E25, and FFV any blend) (Kadiman 2005).

As biomass hydrolysis and sugar fermentation technologies approach commercial viability, advancements in product recovery technologies will be required. For cases in which fermentation products are more volatile than water, recovery by distillation is often the technology of choice. Distillation technologies that will allow the economic recovery of dilute volatile products from streams containing a variety of impurities have been developed and commercially demonstrated. A distillation system separates the bioethanol from water in the liquid mixture.

The first step is to recover the bioethanol in a distillation or beer column, where most of the water remains with the solids part. The product (37% bioethanol) is then concentrated in a rectifying column to a concentration just below the azeotrope (95%). The remaining bottom product is fed to the stripping column to remove additional water, with the bioethanol distillate from stripping being recombined with the feed to the rectifier. The recovery of bioethanol in the distillation columns in the plant is fixed to be 99.6% to reduce bioethanol losses. The use of methanol as a motor fuel received attention during the oil crises of the 1970s due to its availability and low cost. Problems occurred early in the development of gasoline-methanol blends. As a result of its low price some gasoline marketers over blended. Many tests have shown promising results using 85–100% by volume methanol as a transportation fuel in automobiles, trucks, and buses.

Methanol can be used as one possible replacement for conventional motor fuels. During times of oil crisis, methanol has been viewed as a possible large-volume motor fuel substitute. It was often used in the early part of the century to power automobiles before inexpensive gasoline was widely introduced. In the early 1920s, some viewed it as a source of fuel before new techniques were developed to discover and extract oil. Synthetically produced methanol was widely used as a motor fuel in Germany during World War I.

Before modern production technologies were developed in the 1920s, methanol was obtained from wood as a co-product of charcoal production and, for this reason, was commonly known as wood alcohol. Methanol is currently manufactured worldwide by conversion or derived from syngas, natural gas, refinery off-gas, coal or petroleum ($2H_2 + CO \rightarrow CH_3OH$). The chemical composition of syngas from coal and then from natural gas can be identical with the same H_2/CO ratio. A variety of catalysts are capable of causing the conversion, including reduced NiO-based preparations, reduced Cu/ZnO shift preparations, Cu/SiO_2 and Pd/SiO_2, and Pd/ZnO (Takezawa et al. 1987; Iwasa et al. 1993).

Starting from 1980s important progress has been made on evaluating some low-grade oils, oil production wastes and residues as motor fuel (Pryor et al. 1983). But direct usage of vegetable oils causes a number of problems concerning the engine because of their high viscosity and the excessive carbonaceous deposits left in the cylinders and on the injector nozzles. Therefore, chemical conversion of vegetable oils was suggested. In order to lower the viscosities and flash points of vegetable oils the transesterification method was applied and it was reported that the alcoholysis products of soybean, sunflower, rapeseed and used frying oils were proposed as diesel fuel alternatives.

The deregulation of domestic crude oil prices and the formation of OECD have been largely responsible for high fuel prices. The farmer is highly dependent on diesel fuel for crop production. Alternative fuels such as vegetable oils could easily help the petroleum dependence of farmers. Recently, the demand for crude oil has decreased because of conservation practices but ultimately a liquid fuel resource problem exists (Pryor et al. 1983).

Biodiesel is a mixture of methyl esters of long-chain fatty acids like lauric, palmitic, steric, oleic, etc. Typical examples are rapeseed oil, canola oil, soybean oil, sunflower oil, palm oil and its derivatives from vegetable sources. Beef and sheep tallow and poultry oil from animal sources and cooking oil are also the sources of raw materials. The chemistry of conversion to biodiesel is essentially the same. Oil or fat react with methanol or ethanol in the presence of catalyst sodium hydroxide or potassium hydroxide to form biodiesel, (m)ethylesters, and glycerin. Biodiesel is technically competitive with or offers technical advantages compared to conventional petroleum diesel fuel. The biodiesel esters were characterized for their physical and fuel properties including density, viscosity, iodine value, acid value, cloud point, pure point, gross heat of combustion, and volatility. The biodiesel fuels produced slightly lower power and torque, and higher fuel consumption than No. 2 diesel fuel. Biodiesel is better than diesel fuel in terms of sulfur content, flash point, aromatic content and biodegradability. Some technical properties of biodiesels are shown in Table 4.6.

The advantages of biodiesel as diesel fuel are its liquid nature portability, ready availability, renewability, higher combustion efficiency, lower sulfur and aromatic content (Knothe et al. 2006; Demirbas 2008), higher cetane number, and higher biodegradability (Zhang et al. 2003). The main advantages of biodiesel given in the literature include domestic origin, reducing the dependency on imported petroleum, biodegradability, high flash point, and inherent lubricity in the neat form (Knothe et al. 2005). The disadvantages of biodiesel are higher viscosity, lower energy content, higher cloud point and pour point, higher nitrogen oxide (NO_x) emissions, lower engine speed and power, injector coking, engine compatibility, high price, and higher engine wear.

Table 4.6 Some technical properties of biodiesels

Common name	Biodiesel
Common chemical name	Fatty acid (m)ethyl ester
Chemical formula range	C_{14}–C_{24} Methyl esters or C_{15}–C_{25} $H_{28-48}O_2$
Kinematic viscosity range (mm^2/s, at 313 K)	3.3–5.2
Density range (kg/m^3, at 288 K)	860–894
Boiling point range (K)	>475
Flash point range (K)	428–453
Distillation range (K)	470–600
Vapor pressure (mm Hg, at 295 K)	<5
Solubility in water	Insoluble in water
Physical appearance	Light to dark yellow, clear liquid
Odor	Light musty/soapy odor
Biodegradability	Higher biodegradability than petroleum diesel
Reactivity	Stable, but avoid strong oxidizing agents

The term bio-oil is used mainly to refer to liquid fuels from biorenewable feedstocks. Biomass is heated in the absence of oxygen, or partially combusted in a limited oxygen supply, to produce an oil-like liquid, a hydrocarbon-rich gas mixture and a carbon-rich solid residue. Pyrolysis dates back to at least ancient Egyptian times, when tar for caulking boats and certain embalming agents were made by pyrolysis. In the 1980s, researchers found that the pyrolysis liquid yield could be increased using fast pyrolysis, where a biomass feedstock is heated at a rapid rate and the vapors produced also condensed rapidly (Mohan et al. 2006).

In wood-derived pyrolysis oil, specific oxygenated compounds are present in relatively large amounts. A current comprehensive review focuses on the recent developments in the wood/biomass pyrolysis and reports the characteristics of the resulting bio-oils, which are the main products of fast wood pyrolysis (Mohan et al. 2006). Sufficient hydrogen added to the synthesis gas to convert all of the biomass carbon into methanol carbon would more than double the methanol produced from the same biomass base.

The kinematic viscosity of pyrolysis oil varies from as low as 11 cSt to as high as 115 mm^2/s (measured at 313 K) depending on nature of the feedstock, temperature of pyrolysis process, thermal degradation degree and catalytic cracking, the water content of the pyrolysis oil, the amount of light ends that have collected, and the pyrolysis process used. The pyrolysis oils have water contents of typically 15–30 wt.% of the oil mass, which cannot be removed by conventional methods like distillation. Phase separation may partially occur above certain water contents. The water content of pyrolysis oils contributes to their low energy density, lowers the flame temperature of the oils, leads to ignition difficulties, and, when preheating the oil, can lead to premature evaporation of the oil and resultant injection difficulties. The higher heating value (HHV) of pyrolysis oils is below 26 MJ/kg (compared to 42–45 MJ/kg for conventional petroleum fuel oils). In contrast to petroleum oils, which are non-polar and in which water is insoluble, biomass oils are highly polar and can readily absorb over 35% water (Demirbas 2007).

4.8 Biorenewable Gaseous Fuels

The main biorenewable gaseous fuels are biogas, landfill gas, gaseous fuels from pyrolysis and gasification of biomass, gaseous fuels from Fischer–Tropsch synthesis and biohydrogen. There are a number of processes for the converting of biomass into gaseous fuels such as methane or hydrogen. One pathway uses plant and animal wastes in a fermentation process leading to biogas from which the desired fuels can be isolated. This technology is established and in widespread use for waste treatment. Anaerobic digestion of biowastes occurs in the absence of air, the resulting gas called as biogas is a mixture consisting mainly of methane and carbon dioxide. Biogas is a valuable fuel, which is produced in digesters filled with the feedstock like dung or sewage. The digestion is allowed to continue for a period from ten days to a few weeks. A second pathway uses algae and bacteria that have been genetically modified to produce hydrogen directly instead of the conventional biological energy carriers. Finally, high-temperature gasification supplies a crude gas, which may be transformed into hydrogen by a second reaction step. This pathway may offer the highest overall efficiency.

Anaerobic digestion (AD) is a bacterial fermentation process that is sometimes employed in wastewater treatment for sludge degradation and stabilization. This is also the principal process occurring in the decomposition of food wastes and other biomass in landfills. The AD operates without free oxygen and results in a fuel gas called biogas, containing mostly CH_4 and CO_2 but frequently carrying other substances such as moisture, hydrogen sulfide (H_2S), and particulate matter that are generally removed prior to use of the biogas. AD is a biochemical process for converting biogenic solid waste into a stable, humus-like product. Aerobic conversion uses air or oxygen to support the metabolism of the aerobic microorganisms degrading the substrate. Aerobic conversion includes composting and activated sludge wastewater treatment processes. Composting produces useful materials, such as mulch, soil additives and amendments, and fertilizers.

Anaerobic digestion requires attention to the nutritional needs and the maintenance of reasonable temperatures for the facultative and methanogenic bacteria degrading the waste substrates. The carbon/nitrogen (C/N) ratio of the feedstock is especially important. Biogas can be used after appropriate gas cleanup as a fuel for engines, gas turbines, fuel cells, boilers, industrial heaters, other processes, and the manufacturing of chemicals. Anaerobic digestion is also being explored as a route for direct conversion to hydrogen.

Cellulose and hemicelluloses can be hydrolyzed to simple sugars and amino acids that are consumed and transformed by the fermentative bacteria. The lignin is refractory to hydrolysis and generally exits the process undigested. In fact, lignin may be the most recalcitrant naturally produced organic chemical. Lignin polymers are cross-linked carbohydrate structures with molecular weights on the order of 10 000 atomic mass units. As such, lignin can bind with or encapsulate some cellulose making that cellulose unavailable to hydrolysis and digestion. Lignin degradation (or delignification of lignocellulosics) in nature is due principally to aerobic filamentous fungi that decompose the lignin in order to gain access to the cellulose and hemicelluloses.

For anaerobic systems, methane gas is an important product. Depending on the type and nature of the biological components, different yields can be obtained for different biodegradable wastes. For pure cellulose, for example, the biogas product is 50% methane and 50% carbon dioxide. Mixed waste feedstocks yield biogas with methane concentrations of 40–60% (by volume). Fats and oils can yield biogas with 70% methane content.

Anaerobic digestion functions over a wide temperature range from the so-called psychrophilic temperature near 283 K, to extreme thermophilic temperatures above 344 K. The temperature of the reaction has a very strong influence on the anaerobic activity, but there are two optimal temperature ranges in which microbial activity and biogas production rate are highest, the so-called mesophilic and thermophilic ranges. The mesophilic regime is associated with temperatures of about 308 K, the thermophilic regime of about 328 K. Operation at thermophilic temperatures allows for shorter retention time and a higher biogas production rate; however, maintaining the high temperature generally requires an outside heat source because anaerobic bacteria do not generate sufficient heat. Aerobic composting can achieve relatively high temperatures (up to 344 K) without heat addition because reaction rates for aerobic systems are much higher than those for anaerobic systems. If heat is not conducted away from the hot center of a compost pile, then thermochemical reactions can initiate, which can lead to spontaneous combustion if sufficient oxygen reaches the hot areas. Managed compost operations use aeration to provide oxygen to the bacteria but also to transport heat out of the pile. The molecular structure of the biodegradable portion of the waste that contains proteins and carbohydrates is first broken down through hydrolysis. The lipids are converted to volatile fatty acids and amino acids. Carbohydrates and proteins are hydrolyzed to sugars and amino acids. In acetogenesis, acid-forming bacteria use these byproducts to generate intermediary products such as propionate and butyrate. Further microbial action results in the degradation of these intermediary products into hydrogen and acetate. Methanogenic bacteria consume the hydrogen and acetate to produce methane and carbon dioxide.

After the first 6 days of digestion, methane production from manure increased exponentially, after 16 days it reaches a plateau value, and at the end of day 20, the digestion reached the stationary phase. For the wheat straw and the mixtures of manure/straw the rates of digestion are lower than that of manure.

The maximum daily biogas productions are between 4 and 6 days. During a 30-day digestion period, approximately 80–85% of the biogas is produced in the first 15–18 days. This implies that the digester retention time can be designed to 15–18 days instead of 30 days.

For the first 3 days, methane yield is almost 0% and carbon dioxide generation is almost 100%. In this period, digestion occurred as fermentation to carbon dioxide. The yields of methane and carbon dioxide gases are fifty-fifty on day 11. At the end of day 20, the digestion reaches the stationary phase. The methane content of the biogas is in the range of 73 to 79% for the runs, the remainder being principally carbon dioxide. During digestion, the volatile fatty acid concentration is lower and the pH higher. The pH of the slurry with manure increased from 6.4 initially, to 6.9–7.0 at the maximum methane production rate. The pH of the slurry with wheat straw is around 7.0–7.1 at the maximum methane production rate.

Syngas (a mixture of carbon monoxide and hydrogen) can be produced by gasification of biorenewable feedstocks, which is also called as biosyngas. Biosyngas can be converted into a large number of organic compounds that are useful as chemical feedstocks, fuels, and solvents.

To produce biosyngas from biorenewable feedstocks the following procedures are necessary: (a) gasification of the fuel, (b) cleaning of the product gas, (c) usage of the synthesis gas as energy carrier in fuel cells, and (d) usage of the synthesis gas to produce chemicals.

References

AICHE. 1997. Alternative transportation fuels: a comparative analysis. Report by AICHE Government

Relations Committee. AICHE, New York.

Anderson, R.B. 1984. The Fischer–Tropsch synthesis. Academic, New York.

Berry, G.D., Pasternak, A.D., Rambach, G.D., Smith, J.R., Schock, R.N. 1996. Hydrogen as a future transportation fuel. Energy 21:289–303.

Boerrigter, H., den Uil, H. 2002. Green diesel from biomass via FTS: new insights in gas cleaning and process design. Pyrolysis and Gasification of Biomass and Waste, Expert Meeting, 30 Sep–1 Oct, Strasbourg, France.

Cherry, R.S. 2004. A hydrogen utopia? Int J Hydrogen Energy 29:125 –129.

Chigier, N.A. 1981. Energy, combustion and the environment. McGraw-Hill, New York.

Demirbas, A. 2007. Progress and recent trends in biofuels. Prog Energy Combus Sci 33:1–18.

Demirbas, A. 2008. Biofuels sources, biofuel policy, biofuel economy and global biofuel projections. Energy Convers Manage 49:2106–2116.

Demirbas, A. 2009. Future Energy Sources: Part I. Future Energy Source 1:1–95.

Dry, M.E. 2002. The Fischer–Tropsch process: 1950–2000. Catal Today 71:227–241.

Dry, M.E. 2004. Present and future applications of the Fischer–Tropsch process. Appl Catal A General 276:1–3.

Han, S.-K., Shin, H.-S. 2004. Biohydrogen production by anaerobic fermentation of food waste. Int J Hydrogen Energy 29:569–577.

Iwasa, N., Kudo, S., Takahashi, H., Masuda, S., Takezawa, N. 1993. Highly selective supported Pd catalysts for steam reforming of methanol. Catal Lett 19:211–216.

Jager, B. 1998. In: Proceedings of the 5th Natural Gas Conversion Symposium, Taormina, Italy.

Jin, Y., Datye, A.K. 2000. Phase transformations in iron Fischer–Tropsch catalysts during temperature-programmed reduction. J Catal 196:8–17.

Kadiman, O.K. 2005. Crops: beyond foods. In: Proceedings of the 1st International Conference of Crop Security, Malang, Indonesia, 20–23 Sept 2005.

Knothe, G., Krahl, J., Van Gerpen, J. (Eds.) 2005. The Biodiesel Handbook. AOCS, Champaign, IL.

Knothe, G., Sharp, C.A., Ryan, T.W. 2006. Exhaust emissions of biodiesel, petrodiesel, neat methyl esters, and alkanes in a new technology engine. Energy Fuels 20:403–408.

Lang, X., Macdonald, D.G., Hill, G.A. 2001. Recycle bioreactor for bioethanol production from wheat starch II. Fermentation and economics. Energy Sources 23:427–436.

MacLean, H.L., Lave, L.B. 2003. Evaluating automobile fuel/propulsion system technologies. Progress Energy Combus Sci 29:1–69.

Mislavskaya, V.S., Leonow, V.E., Mislavskii, N.O., Ryzhak, I.A. 1982. Conditions of phase stability in a gasoline-methanol-cyclohexanol-water system. Soviet Chem Ind 14:270–276.

Mohan, D., Pittman, C.U., Steele, P.H. 2006. Pyrolysis of wood/biomass for bio-oil: a critical review. Energy Fuels 20:848–889.

Momirlan, M., Veziroglu, T. 1999. Recent directions of world hydrogen production. Renew Sust Energy Rev 3:219–231.

Momirlan, M., Veziroglu, T. 2002. Current status of hydrogen energy. Renew Sustain Energy Rev 6:141–179.

Osten, D.W., Sell, N.J. 1983. Methanol-gasoline blends: blending agents to prevent phase separation. Fuel 62:268–270.

Piel, W.J. 2001. Transportation fuels of the future? Fuel Process Technol 71:167–179.

Plass, J.H., Barbir, F., Miller, H.P., Veziroglu, T.N. 1990. Economics of hydrogen as a fuel for surface transportation. Int J Hydrogen Energy 15:663–668.

Pryor, R.W., Hanna, M.A., Schinstock, J.L., Bashford, L.L. 1983. Soybean oil fuel in a small diesel engine. Trans ASAE 26:333–338.

Spath, P.L., Dayton, D.C. 2003. Preliminary screening: technical and economic assessment of synthesis gas to fuels and chemicals with emphasis on the potential for biomass-derived syngas. NREL/TP-510-34929. National Renewable Energy Laboratory, Golden, CO.

Spath, P.L., Mann, M.K. 2000. Life cycle assessment of hydrogen production via natural gas steam reforming. TP-570-27637, November. National Renewable Energy Laboratory, Golden, CO.

Takezawa, N., Shimokawabe, M., Hiramatsu, H., Sugiura, H., Asakawa, T., Kobayashi, H. 1987. Steam reforming of methanol over Cu/ZrO_2. Role of ZrO_2 support. React Kinet Catal Lett 33:191–196.

Tijmensen, M.J.A., Faaij, A.P.C., Hamelinck, C.N., van Hardeveld, M.R.M. 2002. Exploration of the possibilities for production of Fischer–Tropsch liquids and power via biomass gasification. Biomass Bioenergy 23:129–152.

Vosloo, A.C. 2001. Fischer–Tropsch: a futuristic view. Fuel Proc Technol 71:149–155.

Zhang, Y., Dub, M.A., McLean, D.D., Kates, M. 2003. Biodiesel production from waste cooking oil: 2. Economic assessment and sensitivity analysis. Biores Technol 90:229–240.

Chapter 5
Biomass Fractionation and Valorization

5.1 Introduction

A facility where biomass is converted into value-added products can be referred to as a *biorefinery*. Biorefineries of the future will produce a wide array of products derived from biomass that cannot even be envisioned at this time. In the near term, biorefineries must rely on products that already have a market in today's economy.

Biomass energy is one of humanity's earliest sources of energy particularly in rural areas where it is often the only accessible and affordable source of energy. Worldwide biomass ranks fourth as an energy resource, providing approximately 14% of the world's energy needs. All human and industrial processes produce waste, that is, normally unused and undesirable products of a specific process. Generation and recovery of solid wastes varies dramatically from country to country and deserves special mention. The burning velocity of pulverized biomass fuels is considerably higher than that of coal. The use of biomass fuels provides substantial benefits as far as the environment is concerned. Biomass absorbs carbon dioxide during growth, and emits it during combustion. Utilization of biomass as fuel for power production offers the advantage of a renewable and CO_2-neutral fuel. Although the structural, proximate, and ultimate analysis results of biomass and wastes differ considerably, some properties of the biomass samples such as the hydrogen content, the sulfur content and the ignition temperatures changed in a narrow interval (Demirbas 2004).

Lignocellulosic-containing plants are a composite polymeric material containing primarily hemicelluloses, cellulose, and lignin, each with a wide range of uses and byproducts including sugars, ethanol, and other chemical products. Hemicellulose are very complex polymers composed of various five and six-carbon sugars in a highly branched structure. Hemicelluloses are heteropolymers composed of sugar monomers, including xylose, mannose, glucose, galactose and others, which can also have side chains. Hemicellulose attaches weakly to both cellulose and lignin and fills the intervening spaces. The hemicellulose fraction has potential for being hydrolyzed into five-carbon sugars, which are building blocks for numerous products and industrial chemicals or producing energy by aerobically digesting the sugars to methane. Cellulose, a linear polymer of anhydroglucose units, comprises 40 to 50% of the oven-dry mass of wood. The cellulose fraction has potential use as a fiber or as a feedstock for the production of fuel-grade ethanol. Hemicelluloses tend to be much shorter in length than cellulose, and the molecular structure is slightly branched. Lignin is a completely different polymeric material, being highly cross-linked and having phenolic-like structures as the monomeric base. Lignin is a macromolecule, which consists of alkyl phenols and has a complex three-dimensional structure. Lignin is a complex high molecular weight compound with an even more random structure than hemicellulose. The three most prevalent monomers in lignin, all phenylpropane derivatives, are *p*-coumaryl alcohol, coniferyl alcohol, and sinapyl alcohol (Peterson et al. 2008). The resulting lignin fraction has potential as a source of specialty chemicals, but in the near term, as a fuel for co-firing with coal to produce electricity at the biorefining site. Wood substance can be thought of as a matrix of cellulosic/hemicellulosic cells imbedded in a lignin substrate. It is the lignin that holds the wood cells together and provides the extraordinary composite strength characteristics of a piece of wood. Extractives vary considerably in composition between species. The major extractive categories in woody biomass are aliphatic compounds (fats and waxes), terpenes and terpenoids, and phenolic compounds (Demirbas 2000). Figure 5.1 shows the main components in plant biomass. Typical component analysis of some plant biomass samples are given in Table 5.1 (Demirbas 2009).

Wood biomass involves trees with commercial structure and forest residues not being used in the traditional forest products industries. Available forest residues may appear to be an attractive fuel source. Forest residues are parts of trees such as branches, small-diameter wood, stumps and dead wood as well as undergrowth and low-value species. Principal sources of waste wood are two waste streams: municipal solid waste and construction and demolition waste. The most limiting factors for the use of wood waste as fuel for power generation are transportation costs and its energy content.

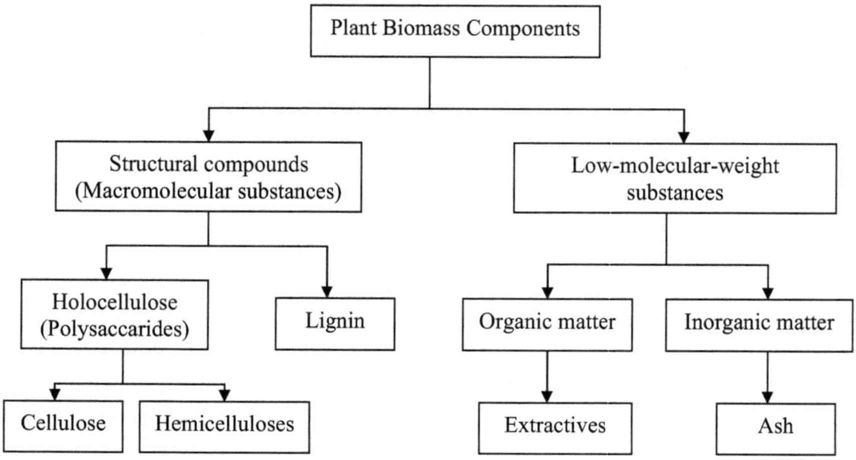

Fig. 5.1 Main components in plant biomass

Table 5.1 Typical component analysis of some plant biomass samples

Biomass sample	Cellulose (%)	Hemicelluloses (%)	Lignin (%)
Apricot stone	22.4	20.8	51.4
Beech wood	44.2	33.5	21.8
Birchwood	40.0	25.7	15.7
Hazelnut shell	25.2	28.2	42.1
Legume straw	28.1	34.1	34.0
Orchard grass	32.0	40.0	4.7
Pine sawdust	43.8	25.2	26.4
Rice straw	34.0	27.2	14.2
Spruce wood	43.0	29.4	27.6
Tea waste	31.2	22.8	40.3
Tobacco stalk	21.3	32.9	30.2

Forest products can be converted into energy through physical, biological and thermochemical processes. Physical processes are grinding, drying, filtration, pressing, extraction, and briquetting. Biological processes are essentially microbic digestion and fermentation. Main thermochemical processes are pyrolysis gasification. Pyrolysis is the process of applying high heat to wood materials in the absence of air or in reduced air. The wood pyrolysis is attractive because forest and industrial wood residues can be readily converted into biofuels and biochemicals. Gasification creates synthetic gas (syngas), which has the potential to produce a wide range of commercial fuels and chemicals. Ethanol has been commercially manufactured from wood carbohydrates biomass feedstocks using acid hydrolysis and subsequent fermentation techniques.

The use of wood energy has many unique qualities that provide environmental benefits. It can help mitigate climate change, reduce acid rain, soil erosion, water pollution and pressure on landfills, provide wildlife habitat, and help maintain forest health through better management. Modern boilers usually produce heat, steam, or electricity. Direct combustion systems vary considerably in their design. The fuel choice makes a difference in the design and efficiency of the combustion system. Direct combustion technology using wood waste as the fuel is very similar to that used for coal. Air toxic emissions during wood combustion were typically very low, and often near or below detection limits, largely as a result of the good air-fuel mixing and high furnace temperatures associated with pulverized coal combustion.

5.2 Biomass Fractionation and Valorization Facilities

The benefits of an integrated biorefinery are numerous because of the diversification in feedstocks and products. There are currently several different levels of integration in biorefineries, which adds to their sustainability, both economically and environmentally. Economic and production advantages increase with the level of integration in the biorefinery. Industrial biorefineries have been identified as the most promising routes to the creation of a bio-based economy. Partial biorefineries already exist in some energy crop, forest-based, and lignocellulosic products facilities. Integrated bio-based industries use a variety of different technologies to produce chemicals, biofuels, food and feed ingredients, biomaterials and power from biomass raw materials. These systems can be improved through better utilization of agricultural residues and solid wastes, and optimization of total value-added products.

Biomass energy (energy from wood, agricultural residues, animal wastes and other organic wastes) currently represents approximately 14% of world final energy consumption. Worldwide biomass ranks fourth as an energy resource, and biomass is the most important source of energy in developing nations, providing around 35% of their energy (McGowan 1991; Hall et al. 1992).

Wood-related industries and households consume the most biomass energy. The lumber, pulp and paper industries burn their own wood wastes in large furnaces and boilers to supply 60% of the energy needed to run the factories. Biomass includes 60% wood and 40% non-wood materials. Table 5.2 shows percentages of world renewable energy sources. Table 5.3 shows the percentage share of each renewable energy source in 1995 (Demirbas 2004). As seen in Table 5.3, the percentage share of biomass was 62.1% of total renewable energy sources in 1995.

Wood biomass involves trees with commercial structure and forest residues not being used in the traditional forest products industries. Forest residues typically refer to those parts of trees such as treetops, branches, small-diameter wood, stumps and dead wood as well as undergrowth and low-value species. The conversion of wood to biofuels and biochemicals has long been a goal of the forest and products industry. Forest residues alone count for some 50% of the total forest biomass and are currently left in the forest to rot (Demirbas 2001).

The opportunity to combine fuelwood production with effluent disposal has many potential environmental benefits. The economic feasibility of woody biomass plantations is difficult to justify at present but if the full cost of externalities such as CO_2 emissions are to be applied in the future, than acceptable payback periods for boiler conversions, as well as handling and storage facilities could be achieved.

Fractionation refers to the conversion of wood into its constituent components (cellulose, hemicelluloses, and lignin). Other processes include steam explosion,

Table 5.2 Percentages of world renewable energy sources

Renewable energy source	Percentage
Hydro	34
Wood biomass in households	30
Non-wood biomass	19
Wood biomass in industry	13
Geothermal, wind, solar and other renewables	4

Table 5.3 Percentage share of each renewable energy source in 1995

Resource	Percentage
Biomass	62.1
Wood in household	30.3
Wood in industry	12.7
Power station	11.8
Municipal solid waste	3.7
District heating	1.8
Biogas	1.0
Liquid fuels	0.8
Hydro	33.6
Geothermal	3.2
Wind	0.7
Solar	0.4

aqueous separation and hot water systems. Commercial products of biomass fractionation include levulinic acid, xylitol, and alcohols. Figure 5.2 shows the fractionation of wood and chemicals from wood. Figure 5.3 shows the separation of plant biomass into cellulose product and lignin stream.

Main fractionation chemicals from wood ingredients are:

Dissociation of cell components → Lignin fragment + Oligosaccharides + Cellulose (5.1)

Hydrolysis of cellulose (saccharification) → Glucose (5.2)

Conversion of glucose (fermentation) → Ethanol + Lactic acid (5.3)

Chemical degradation of cellulose → Levulinic acid + Xylitol (5.4)

Chemical degradation of lignin → Phenolic products (5.5)

Wood fuel is a renewable energy source and its importance will increase in future. Three main determinants of the costs of operating and constructing a wood-fired power plant of a given size exist: (1) the availability of the required fuel, (2) the delivered fuel prices, and (3) the financing and construction of the desired power plant (Joutz 1992). Most previous research has focused on the feasibility of single-source fuel plants. One popular approach has been to examine the feasibility of short-rotation intensive-cultivation plantations (Jennergren and Thornqvist 1998).

Main current wood valorization technologies are:

1. Manufacturing of wooden materials
   (miner and postal poles, crosstie, constructors)
2. Direct combustion as fuelwood
3. Pulp and paper making

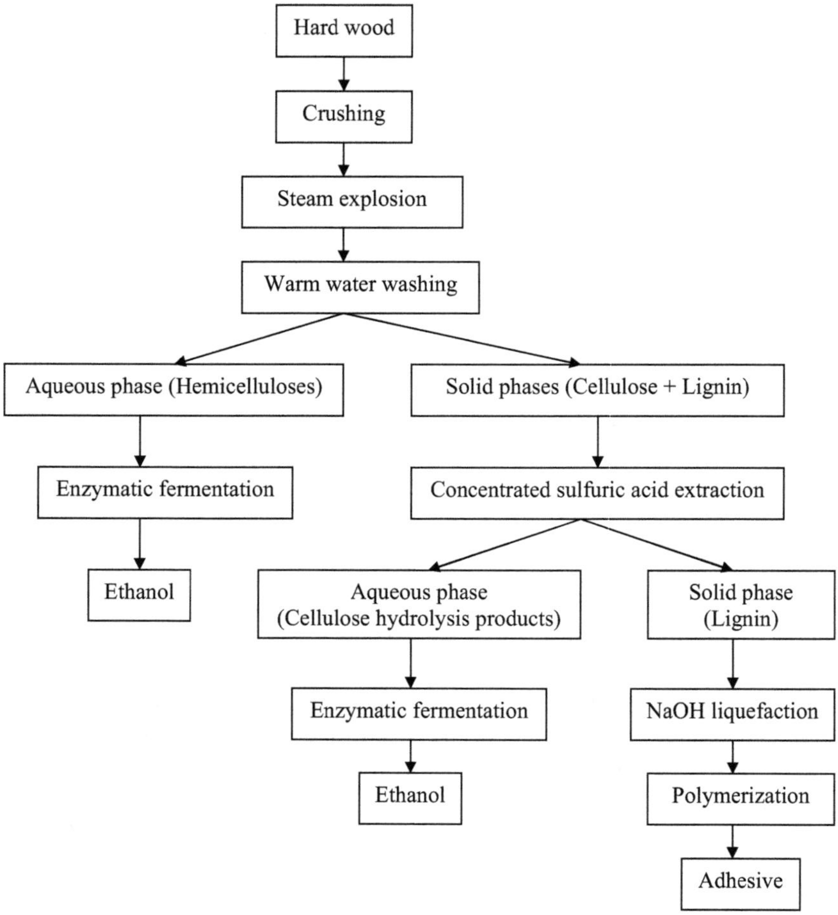

Fig. 5.2 Fractionation of wood and chemicals from wood

4. Bio-oil by pyrolysis
5. Synthesis gas by gasification
6. Char from wood by slow carbonization
7. Sugar from wood by hydrolysis
8. Ethanol by hydrolysis and fermentation
9. Adhesive from lignin by alkali liquefaction and polymerization
10. Chemicals by Soxhlet and supercritical fluid extractions

The conversion of wood into chemicals for the production of most of our synthetic plastics, fibers, and rubbers is technically feasible. Synthetic oils from liquefaction of wood might serve as feedstocks for cracking to chemicals in the same way that crude oil is presently used (Goldstein 1975).

A current comprehensive review focuses on the recent developments in the wood/biomass pyrolysis and reports the characteristics of the resulting bio-oils, which are

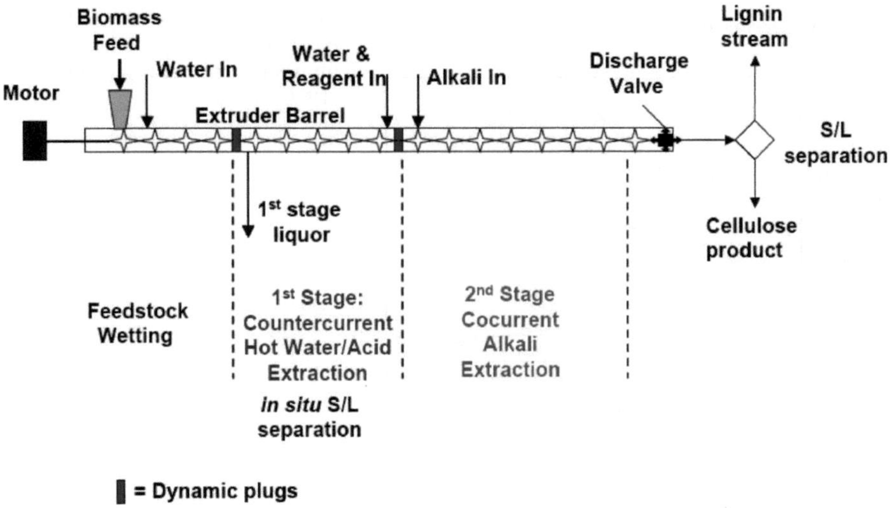

Fig. 5.3 Separation of plant biomass into cellulose product and lignin stream

the main products of fast pyrolysis of wood (Mohan et al. 2006). Fast pyrolysis utilizes wood biomass to produce a product that is used both as an energy source and a feedstock for chemical production. Considerable efforts have been made to convert wood biomass to liquid fuels and chemicals since the oil crisis in 1970s.

5.3 Physical Valorization Facilities of Biomass

Forests are a principal global economic resource as well as an ecological resource. Forests have played an important role in the development of human societies. The prime direct or marketable product of most forests is wood for use as timber, fuelwood, pulp and paper, providing some 3.4 billion cubic meters of timber-equivalent a year globally. Asia and Africa use 75% of global wood fuels.

The lumber, plywood, pulp, and paper industries burn their own wood residues in large furnaces and boilers to supply 60% of the energy needed to run factories (Demirbas 2003). Figure 5.4 shows the use of world wood products such as lumber, plywood, paper, and paperboard between 1970 and 2005. The availability of fuelwood from the forest is continually declining at an increasing rate due to indiscriminate deforestation and slow regeneration, as well as afforestation (Jain and Singh 1999).

The fuelwoods generally used by local people were identified and analyzed quantitatively to select a few species with the best fuelwood characteristics so that plantation on non-agricultural lands could be undertaken (Jain 1992).

The reduction of particle size and moisture content, together with the most appropriate storage and handling systems are necessary for an efficiently operated wood waste combustion system. Size reduction may be carried out in several stages

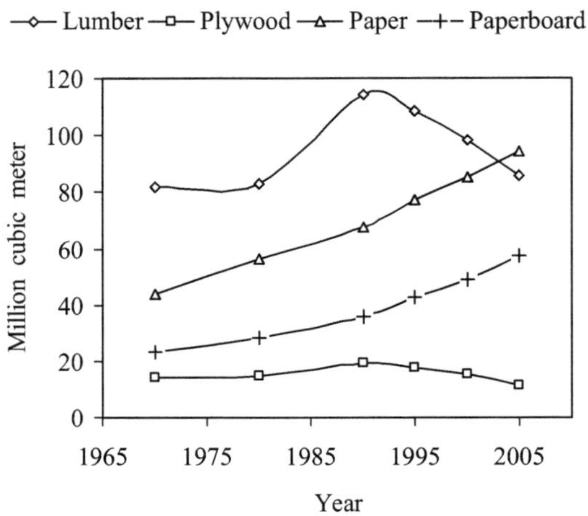

Fig. 5.4 Use of world wood products including lumber, plywood, paper, and paperboard (1970–2005)

in a hog or attrition mill, with screening before and in between. The moisture in residues may be reduced either by mechanical pressing, air drying or the use of hot air dryers, or a combination of all three. Generally slabs, edgings, peeler cores, veneer waste and trimmings would be transported by mechanical conveyors.

Collection and handling and transport costs are critical factors in the use of forest residues. Although the heat produced from wood wastes is less than that from oil or gas, its cost compared to fossil fuels makes it an attractive source of readily available heat or heat and power. The most effective utilization of wood wastes, particularly in the sawmilling and plywood industry, plays an important role in energy efficient production. Table 5.4 shows the sources of available forest and wood manufacturing residues. Wood preparation involves the conversion of roundwood logs into a form suitable for pulping and includes processes for debarking, chipping, screening, handling, and storage.

There are two waste streams: municipal solid waste (MSW) and construction and demolition waste (C&DW). MSW is waste from residential, commercial, institutional, and industrial sources. Each generates distinctly different types of wood waste, with differing degrees and levels of recyclability. The primary components of wood waste are used lumber, shipping pallets, trees, branches, and other wood

Table 5.4 Sources of available forest and wood manufacturing residues

Source of residue	Type of residue
Forest operations	Branches, bark chips, leaves/needles, stumps, roots, and sawdust
Lumber production	Bark, sawdust, clippings, split wood
Sawmilling	Bark, trimmings, split wood, sawdust, planer shavings, sander dust
Plywood production	Bark, sawdust, veneer clippings and wastes, panel trim, sander dust
Paper production	Slab chips, pulping reject, sawdust, clippings
Paperboard production	Bark, sawdust, screening fines, panel trim, sander dust

debris from construction and demolition clearing and grubbing activities. Construction and demolition (C&D) waste is defined as solid, largely inert waste resulting from the construction, repair, demolition, or razing of buildings, roads and other structures. The term also includes debris from the clearing of land for construction. Other sources of waste wood include chemically treated wood from railroad ties, telephone and utility poles, miner poles, crosstie, constructors, pier and dock timbers, untreated wood from logging and silvicultural operations, and industrial waste wood outside the MSW and C&DW streams. Chemical treatments and costs of collection make much of this material difficult to recover.

The wood waste preparation process generally involves hogging, dewatering, screening, size reduction, bulk storage, blending and drying prior to combustion so as to ensure a reliable and consistent supply of quality fuel to the burners. The handling, treatment and storage of wood waste fuels are considerably more costly and troublesome than that required for traditional fossil fuels.

5.3.1 Briquetting of Particulate Wood Waste

Briquetting may be considered as a good idea to utilize low-grade combustible materials obtained from waste materials. Briquettes or logs are generally formed by forcing dry sawdust or shavings through a split cylindrical die using a hydraulic ram. The exerted pressure, of some 1200 kg/cm^2, and the resultant heat generated bonds the wood particles into "logs." Various particulate combustible wastes such as coal dust, chopped polyethylene, sawdust, milled bark, chopped car tires, and chopped horticultural wastes have been considered for bonding with paper fibers (Liu 1999). Since the economic feasibility strongly depends on local conditions, only general figures will be presented to set the level of costs and show the market situation. It is possible to get a better economic return using wood powder instead of less upgraded wood fuels. The production of pellets or logs involves the reduction of wood waste to the size of sawdust, which is then dried to approximately 12% moisture content, before being extruded in specially adapted agricultural pellet mills to form pellets of some 6 to 18 mm diameter and 15 to 30 mm long, with a density in the range of 950 to 1300 kg/m^3.

Densification of biomass waste materials makes the material denser and easier to handle, thus, reducing the cost of transportation and handling. Densification can also increase the heating value per unit volume, making the biomass fuel more compatible with coal and more efficient in combustion. However, the densification of biomass waste materials, such as sawdust, wood chips, tree trimmings, waste paper and the combustible part of municipal solid waste, is usually patterned after the century-old mature technology of coal briquetting (Demirbas et al. 2004).

The briquetting of char improves and provides more efficient use of biomass-based energy resources such as wood and agricultural wastes. The char briquettes that are sold on the commercial market are typically made from a binder and filler. The char is crushed into fines and passed through a variety of screens to make sure the particle size is small enough. A binder, typically starch, is added to the

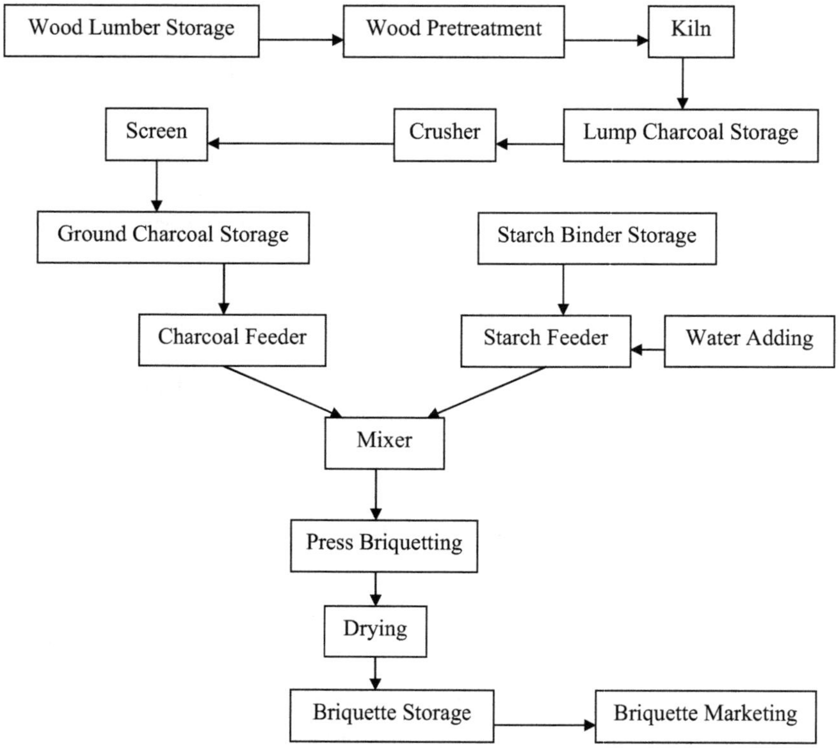

Fig. 5.5 Flow diagram for char briquette production

fines, as well as water. Starch is preferred over other alternatives (wax and wood pitch) because of its economical price and availability. As the material flows to the mixer, metered amounts of about 5% of binder (potato or corn starch) with water are added. Char compromises 75% of the briquette mixture, while water and starch compose 20% and 5%, respectively (Moscowitz 1978). Figure 5.5 depicts the flow diagram for char briquette production. The press for briquetting must be well designed, strongly built and capable of agglomerating the mixture of charcoal and binder sufficiently for it to be handled through the drying process.

Manufacturing of briquettes from raw material may be either an integral part of a charcoal producing facility, or an independent operation, with charcoal being received as raw material. Briquettes are a processed biomass fuel that can be burned as an alternative to wood or charcoal for heat energy. Often they are used for cooking.

Wood has very low sulfur content and so does charcoal. Continuous production of charcoal is more amenable to emission control than batch production because emission composition and flow rate are relatively constant. Briquette production also reduced the impacts on health by providing a cleaner burning fuel. If coaling temperatures are too low, excessive amounts of volatiles will remain in the charcoal and cause heavy smoke when it burns.

5.4 Chemical Valorization Facilities of Biomass

Main wood constituent are hemicelluloses, cellulose, and lignin. Hemicelluloses of the xylan family from plant polysaccharides were valorized in order to produce hydrophobic films by the esterification process (Moine et al. 2005). Cellulose is a biopolymer of anhydroglucose units. Thermal degradation of cellulose proceeds through two types of reaction: a gradual degradation, decomposition and charring on heating at lower temperatures, and a rapid volatilization accompanied by the formation of levoglucosan on pyrolysis at higher temperatures. The initial degradation reactions include depolymerization, hydrolysis, oxidation, dehydration, and decarboxylation (Shafizadeh and Stevenson 1982).

Pyrolysis is the thermochemical process that converts organic materials into usable fuels. Pyrolysis produces energy fuels with high fuel-to-feed ratios, making it the most efficient process for biomass conversion and the method most capable of competing with and eventually replacing non-renewable fossil fuel resources. The wood pyrolysis is attractive because forest and industrial wood residues can be readily converted into liquid products. These liquids, as crude bio-oil or slurry of charcoal of water or oil, have advantages in transport, storage, combustion, retrofitting and flexibility in production and marketing (Demirbas 2007).

In the first step of pyrolysis of carbohydrates dehydration occurs and at low temperatures dehydration predominates. Dehydration is also known as char forming reaction. Between 550 and 675 K volatile products, tar, and char are formed. The volatile products are CO_2, CO, H_2O, acetals, furfural, aldehydes and ketones. Levoglucosan is the principle component in tar.

The holocellulose leads to the formation of furans and carbohydrates, whereas the lignin generates phenolics. Syringols and guaiacols, originating from primary degradation of lignin, reach their maximum levels in the bio-oils during pyrolyses at about 750 to 800 K. In contrast, secondary degradations continuously increase production of phenols with increasing temperature. The pyrolysis of lignin is endothermic between 375 and 450 K. Maximum pyrolytic conversion rate occurs between 625 and 725 K. Char, liquid including methanol, methoxy compounds, acetic acid, phenolic compounds and acetone, and gas including CO, methane, CO_2 and ethane are formed.

Lignin is a highly phenolic functional polymer. Lignins can be modified via sodium hydroxide liquefaction. Functionalization of modified lignin leads to selective structural modifications (Funaoka et al. 1995). Lignin modification products can be converted to adhesives using phenol formaldehyde (PF) and polymeric methylene diphenyl diisocyanate (pMDI) adhesives. PF and pMDI are two common structural adhesives used for various composite wood products. Low-cost wood adhesives can be obtained from the wastes of particleboard and polywood plants.

Phenolics, including simple phenols (mostly phenolic acids), flavonoids and anthocyanins, are hydrophilic compounds with anti-oxidant activity in vitro (Ju and Bramlage 1999). Phenolic phytochemicals are important aromatic secondary metabolites in plants, many of which are commonly substituted by sugar moieties such as glucose, arabinose, xylose, rhamnose and galactose (Kim et al. 2003). Many phenolic phytochemicals have anti-oxidative, anti-carcinogenic, anti-microbial, anti-allergic, anti-mutagenic and anti-inflammatory activities (Kawaii et al. 1999; Eberhardt et al. 2000; Kim et al. 2000).

To obtain ethanol, wood biomass can be converted by hydrolysis into sugars and subsequently fermentation of sugars. Fermentation is a well-understood biological process in which enzymes produced by microorganisms catalyze chemical reactions that convert simple sugars into alcohol. Ethanol is the primary product of fermentation, though the process can be tailored to produce various quantities of other alcohols. One specific yeast (*Saccharomyces cerevisiae* also known as baker's yeast) is frequently used to ferment glucose to ethanol.

There are three basic types of sugar obtaining processes – acid hydrolysis, enzymatic hydrolysis, and thermochemical – with variations for each. The most common is acid hydrolysis. Virtually any acid can be used; however, sulfuric acid is most commonly used since it is usually the least expensive.

There are two basic types of acid processes: dilute acid and concentrated acid, each with their own variations. Dilute acid processes are conducted under high temperature and pressure, and have reaction times in the range of seconds or minutes, which facilitates continuous processing. Most dilute acid processes are limited to a sugar recovery efficiency of around 50%. The concentrated acid process uses relatively mild temperatures and the only pressures involved are usually only those created by pumping materials from vessel to vessel. The primary advantage of the concentrated process is the high sugar recovery efficiency, which can be on the order of over 90% of both hemicelluloses and cellulose sugars.

There are two bioethanol production processes that currently employ thermochemical reactions in their processes. The first system is actually a hybrid thermochemical and biological system. A genetically engineered microorganism that is capable of converting the synthesis gas is introduced into the fermentation vats under specific process conditions allowing bioethanol to ferment (Thomson 2006).

The second thermochemical bioethanol production process does not use any microorganisms. In this process, biomass materials are first thermochemically gasified and the synthesis gas passed through a reactor containing catalysts, which cause the gas to be converted into bioethanol. Numerous efforts have been made since then to develop commercially viable thermochemical-to-bioethanol processes. Bioethanol yields up to 50% have been obtained using synthesis gas-to-bioethanol processes. Some processes that first produce methanol and then use catalytic shifts to produce bioethanol have obtained bioethanol yields in the range of 80%. Unfortunately, like the other processes, finding a cost-effective all-thermochemical process has been difficult (Badger 2002; Thomson 2006). Table 5.5 shows an overview of conversion routes of plant materials to biofuels (Balat 2009).

Thermochemical processing options appear to be more promising than biological options for the conversion of the lignin fraction of cellulosic biomass. Biological routes of conversion, which can have a detrimental effect on enzymatic hydrolysis, also serve as a source of process energy and potential co-products that have important benefits in a life-cycle context (Lynd et al. 2005).

Gasification is carried out at high temperatures in order to optimize the gas production. Most wood gasification systems utilize air or oxygen in partial oxidation or combustion processes. The resulting gas, known as producer gas, is a mixture of

Table 5.5 Overview of conversion routes of plant materials to biofuels

Plant material	Conversion route	Primarily product	Treatment	Products
Ligno-cellulosic biomass	Flash pyrolysis	Bio-oil	Hydrotreating and refining	C_xH_x, diesel fuel, chemicals, oxygenates, hydrogen
	Gasification	Syngas	Water–gas shift + separation	Hydrogen
			Catalyzed synthesis	Methanol, dimethyl ether, FT diesel, C_xH_x, SNG (CH_4)
	Hydrolysis	Sugar	Fermentation	Bioethanol
	Hydrothermal liquefaction	Bio-oil	Hydrotreating and refining	C_xH_x, diesel fuel, chemicals
	Anaerobic digestion	Biogas	Purification	SNG (CH_4)
Sugar and starch crops	Milling and hydrolysis	Sugar	Fermentation	Bioethanol
Oil plants	Pressing or extraction	Vegetable oil	Esterification	Biodiesel
			Pyrolysis	Bio-oil, diesel fuel, gasoline

carbon monoxide, hydrogen and methane, together with carbon dioxide and nitrogen.

When wood is rapidly heated in a reduced oxygen environment, the feeding material does not combust but rather becomes a synthetic gas (syngas), a combination of hydrogen (H_2) and carbon monoxide (CO). Syngas can be combusted directly in a boiler or used in a gas turbine to generate electricity. The hydrogen from syngas can be isolated and used to power fuel cells. Syngas has the potential to produce to produce a wide range of commercial fuels and chemicals, including synthetic diesel, methanol and lower carbon alcohols, acetic acid, dimethyl ether, etc. (Demirbas 2000).

Commercial gasifiers are available in a range of size and types, and run on a variety of fuels, including wood, charcoal, coconut shells and rice husks. Power output is determined by the economic supply of biomass, which is limited to 80 MW in most regions. Cellulosic wood materials are first thermochemically gasified and the synthesis gas (a mixture of hydrogen and carbon monoxide) bubbled through specially designed fermenters (Badger 2002).

5.5 Reuse of Wood Wastes for Energy Generation

Characteristics influencing wood combustion are (a) particle size and specific gravity, (b) ash content, (c) moisture content, (d) extractive content, (e) element (C,

Table 5.6 Mean elemental analyses of biomass feedstocks including standard deviations (% ash-free basis)

Biomass	C	H	N	O
Wood	52 ±4	6.0 ±0.5	0.3 ±0.3	42 ±4
Agricultural	50 ±1	6.2 ±0.1	0.9 ±0.4	43 ±0.9

H, O and N) content, and (f) structural constituent (cellulose, hemicelluloses and lignin) content (Baker 1982). Mean analyses of biomass feedstocks including standard deviations are given in Table 5.6. There are considerable differences between elemental analyses of wood and agricultural feedstocks.

Figure 5.6 shows the simple wood combustion model. Wood fuels never burn directly: wood fuels are thermally degradable and under the influence of a sufficiently strong energy source they break down into a mixture of volatiles and carbonaceous char. The two modes of combustion (solid char and gaseous volatiles) have completely different chemical mechanisms and kinetics. Biomass combustion is the oldest and the most mature technique of biomass to energy conversion but there lies a great challenge in developing and improving new, more efficient, and environmentally more compatible small and large-scale systems. Fuel properties for the combustion analysis of biomass can be conveniently grouped into physical, chemical, thermal, and mineral properties. Important chemical properties for combustion are the ultimate analysis, proximate analysis, analysis of pyrolysis products, higher heating value, heat of pyrolysis, heating value of the volatiles, and heating value of the char (Kamira 2003). In general fuel properties can be classified as macroscopic or microscopic. Fuel characteristics such as ultimate analysis, heating value, moisture content, particle size, bulk density, and ash fusion temperature of biomass have been reviewed (Baker 1982).

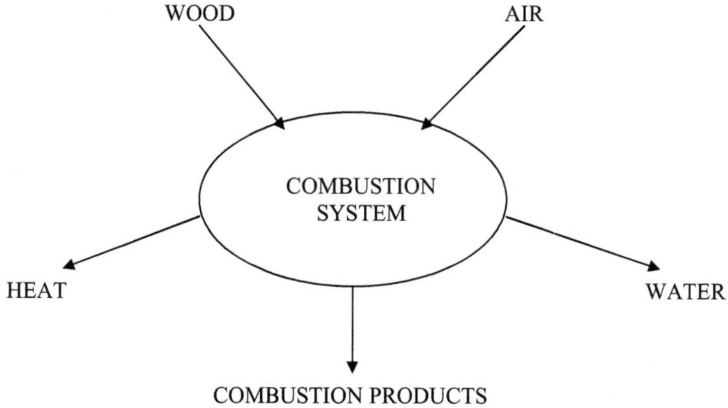

Fig. 5.6 A simple model for wood combustion

Main combustion reactions are:

$$\text{Non-reacting wood} \rightarrow \text{Heat, drying} \rightarrow \text{Degradation (pyrolysis} \rightarrow \text{volatiles)} \rightarrow \text{Pre-combustion reactions} \rightarrow \text{Primary gas phase combustion} \rightarrow \text{Secondary combustion} \rightarrow \text{Effluent stack gas} \quad (5.6)$$

Particle size of biomass should be as much as 0.6 cm, sometimes more, in a profitable combustion process. Biomass is much less dense and has significantly higher aspect ratios than coal. It is also much more difficult to reduce to small sizes. Particle size of wood waste plays an important role in influencing the combustion plant.

Ash or inorganic materials in plants depend on the type of the plant and the soil contamination in which the plant grows. On average wood contains about 0.5% ash. Ash contents of hard and softwoods are about 0.5 and 0.4%, respectively. Insoluble compounds act as a heat sink in the same way as moisture, lowering combustion efficiency, but soluble ionic compounds can have a catalytic effect on the pyrolysis and combustion of the fuel. The presence of inorganic compounds favors the formation of char (Demirbas 1998). Ash content is an important parameter directly affecting the heating value. High ash content of a plant part makes it less desirable as fuel.

Moisture in biomass generally decreases its heating value. Moisture in biomass is stored in spaces within the dead cells and within the cell walls. Green wood contains moisture in two forms: as free water in the cell cavities and as absorbed water in the capillaries of the cell walls. When green wood begins to lose water, the cell walls remain saturated until the free water has evaporated. When the fuel is dried the stored moisture equilibrates with the ambient relative humidity. Equilibrium is usually about 20% in air-dried fuel.

Wood generally has a moisture content of approximately 50 to 55%, although the amount varies according to species, age and the portion of the tree from which it originated, i. e., branches, trunk, etc. Moisture percentage of the wood species varied from 41.27 to 70.20% (Demirbas 2003). Moisture content is a major determinant in the heating value of wood waste, when the moisture content rises to 60%, burning of the wood residues become difficult as its heating value drops dramatically. Heating value of a wood fuel decreases with increasing of moisture content of the wood. Moisture content varies from one tree part to another. It is often the lowest in the stem and increases toward the roots and the crown (Demirbas 2005).

The presence of water in biomass influences its behavior during pyrolysis and affects the physical properties and quality of the pyrolysis liquid. The results obtained show that for higher initial moisture contents the maximum liquid yield on a dry feed basis occurs at lower pyrolysis temperatures between 691 and 702 K (Demirbas 2004).

The heat content, which is a very important factor affecting utilization of any material as a fuel, is affected by the proportion of combustible organic components (called as extractives) present in it (Kataki and Konwer 2001). The HHVs of the extractive-free plant parts were found to be lower than those of the unextracted parts, which indicate a likely positive contribution of extractives towards the increase of HHV. Extractive content is an important parameter directly affecting the heating value. High extractive content of a plant part makes it desirable as fuel. Extractives raised the higher heating values of the wood fuels (Demirbas 2005).

Both the chemical and the physical composition of the fuel are important determining factors in the characteristics of combustion. Biomass can be analyzed by breaking it down into structural components (known as proximate analysis) or into chemical elements (known as ultimate analysis).

The heat content is related to the oxidation state of the natural fuels in which carbon atoms generally dominate and overshadow small variations of hydrogen content. On the basis of literature values for different species of wood, Tillman (1978) also found a linear relationship between HHV and carbon content.

Wood fuels are composed of biopolymers that consist of various types of cells and the cell walls are built of cellulose, hemicelluloses and lignin. HHVs of biomass fuels increase as lignin contents increase.

When evaluating the properties of a combustible material with respect to its use as a fuel, the heating value or energy value is one of the most important factors, which indicates the amount of thermal energy that may be obtained by combusting one mass unit of the material. The heating value of wood depends very much on the species and the part of the tree being used, and varies between 17 to 23 MJ/kg for bone-dry wood; generally softwoods have higher caloric values than hardwoods. For example resinous wood bark, with a high gum and resin content, tends to have a higher value than wood.

The standard measure of the energy content of a fuel is its heating value, sometimes called the calorific value or heat of combustion. The higher heating value at constant pressure measures the enthalpy change of combustion with water condensed. There have been many attempts at correlating the heating value with the composition. Cellulose has a smaller heating value than lignin because of its higher degree of oxidation. Other compounds, such as HC in the fuel with lower degrees of oxidation, tend to raise the heating value of the biomass (Demirbas 2001).

In earlier work (Demirbas 1997), formulae were developed for estimating the HHVs of various lignocellulosic materials, using their ultimate analysis data. The relation between the observed HHV and C, H and O contents of the samples (wt.%) was investigated. Thus the HHV (MJ/kg) of lignocellulosic materials including C, H and O can be calculated from Eq. 5.7:

$$\text{HHV} = 0.335\,\text{C} + 1.423\,\text{H} - 0.154\,\text{O} \tag{5.7}$$

5.6 Biochemical Valorization Facilities of Biomass

Biochemical conversion proceeds at lower temperatures and lower reaction rates can offer high selectivity for products. Higher moisture feedstocks are generally good candidates for biochemical processes. Ethanol production is a biochemical conversion technology used to produce energy from biomass. Many highly efficient biochemical conversion technologies have developed in nature to break down the molecules of which biomass is composed. For ethanol production, biochemical conversion researchers have focused on a process model of dilute acid hydrolysis of hemicelluloses followed by enzymatic hydrolysis of cellulose. Biodiesel production is a biochemical conversion technology used to produce energy from oilseed crops (Shieh et al. 2003; Du et al. 2004; Noureddini et al. 2005).

Ethanol is the most widely used liquid biofuel. It is an alcohol and is fermented from sugars, starches or from cellulosic biomass. Cellulosic materials can be used to produce bioethanol. Bioethanol represents an important, renewable liquid fuel for motor vehicles. Production of bioethanol from biomass is one way to reduce both the consumption of crude oil and environmental pollution. Conversion technologies for producing ethanol from cellulosic biomass resources such as forest materials, agricultural residues and urban wastes are under development and have not yet been demonstrated commercially. In order to produce bioethanol from cellulosic biomass, a pretreatment process is used to reduce the sample size, break down the hemicelluloses to sugars, and open up the structure of the cellulose component. The cellulose portion is hydrolyzed by acids or enzymes into glucose sugar that is fermented to bioethanol. The sugars from the hemicelluloses are also fermented to bioethanol. The use of bioethanol as a motor fuel has as long a history as the car itself. It began with the use of ethanol in the internal combustion engine (Demirbas 2008).

Fermentation feedstocks require pretreatment by chemical, physical, or biological means to open up the structure of biomass and reduce the complex carbohydrates to simple sugars. This set of pretreatments is often referred to as hydrolysis. The resulting sugars can then be fermented by the yeast and bacteria employed in the process. Feedstocks high in starch and sugar are most easily hydrolyzed. Cellulosic feedstocks, including the major fraction of organics in MSW, are more difficult to hydrolyze, requiring more extensive pretreatment.

Hydrolysis is often used to pretreat lignocellulosic feedstocks to break down the cellulose and hemicelluloses from the lignocellulose and break down the compounds into simple sugars. Hydrolysis can be catalyzed by use of acids (either strong or weak), enzymes, and/or hydrothermal means, the latter including hot water and supercritical methods. Figure 5.7 shows the block diagram of the enzymatic hydrolysis process.

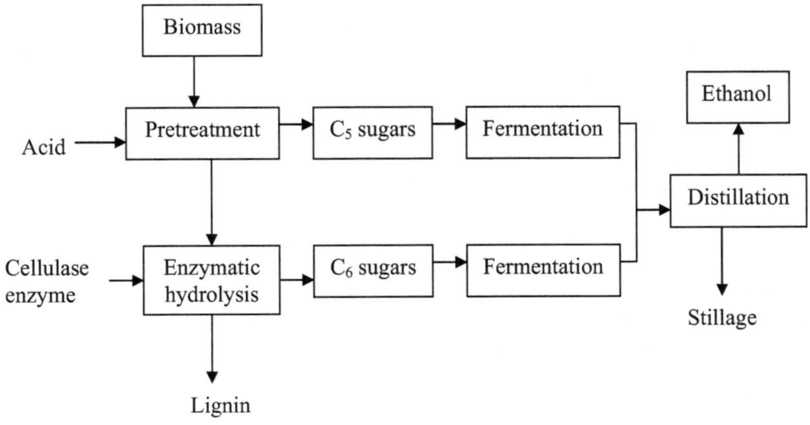

Fig. 5.7 Enzymatic hydrolysis process

Fermentation is generally used industrially to convert substrates such as glucose to ethanol for use in beverage, fuel, and chemical applications and to other chemicals (e.g., lactic acid used in producing renewable plastics) and products (e.g., enzymes for detergents). Strictly speaking, fermentation is an enzymatically controlled anaerobic process although the term is sometimes more loosely applied to include aerobic processing as well.

References

Badger, P.C. 2002. Ethanol from cellulose: A general review. In: Trends in new crops and new uses, J. Janick, J., Whipkey A. (Eds.). ASHS, Alexandria, VA, pp.17–21.
Balat, M. 2009. New biofuel production technologies. Energy Edu Sci Technol Part A 22:147–161.
Baker, A.J. 1982. Wood fuel properties and fuel products from woods. In: Proceedings of Fuelwood, Management and Utilisation Seminar, 9–11 Nov 1982, Michigan State University, East Lansing, MI.
Demirbas, A. 1997. Calculation of higher heating values of biomass fuels. Fuel 76:431–434.
Demirbas, A. 1998. Determination of combustion heat of fuels by using non-calorimetric experimental data. Energy Edu Sci Technol 1:7–12.
Demirbas, A. 2000. Mechanisms of liquefaction and pyrolysis reactions of biomass. Energy Convers Manage 41:633–646.
Demirbas, A. 2001. Biomass resource facilities and biomass conversion processing for fuel and chemicals. Energy Convers Manage 42:1357–1378.
Demirbas, A. 2003. Fuelwood characteristics of six indigenous wood species from Eastern Black Sea Region. Energy Sources 25:309–316.
Demirbas, A. 2004. Global energy sources, energy usage and future developments. Energy Sources 26:191–204.
Demirbas, A. 2005. Potential applications of renewable energy sources, biomass combustion problems in boiler power systems and combustion related environmental issues. Prog Energy Combus Sci 31:171–192.
Demirbas, A. 2007. The influence of temperature on the yields of compounds existing in bio-oils obtained from biomass samples via pyrolysis. Fuel Proc Technol 88:591–598.
Demirbas, A. 2008. The importance of bioethanol and biodiesel from biomass. Energy Sources Part B 3:177–185.
Demirbas, A. 2009. Biofuels: securing the planet's future energy needs. Springer, London.
Demirbas, A., Demirbas-Sahin, A., Demirbas, A.H. 2004. Briquetting properties of biomass waste materials. Energy Sources 26:83–91.
Du, W., Xu, Y., Liu, D., Zeng, J. 2004. Comparative study on lipase-catalyzed transformation of soybean oil for biodiesel production with different acyl acceptors. J Molecular Catal B Enzym 30:125–129.
Eberhardt, M.V., Lee, C.Y., Liu, R.H. 2000. Antioxidant activity of fresh apples. Nature 405:903–904.
Funaoka, M., Matsubara, M., Seki, N., Fukatsu, S. 1995. Conversion of native lignin to a highly phenolic functional polymer and its separation from lignocellulosics. Biotechnol Bioeng 46:545–552.
Goldstein, I.S. 1975. Potential for converting wood into plastics. Chemicals from wood may regain importance as the cost of petroleum continues to rise. Science 189:847–852.
Hall, D.O., Rosillo-Calle, F., de Groot, P. 1992. Biomass energy lessons from case studies in developing countries. Energy Policy January:62–73.
Jennergren, L.P., Thornqvist, T. 1998. An economic model for evaluating peat and logging residues as alternative fuels for a heating plant. Biomass 16:161–172.

Jain, R.K. 1992. Fuelwood characteristics of certain hardwood and softwood tree species of India. Biores Technol 41:129–133.

Jain, R.K., Singh, B. 1999. Fuelwood characteristics of selected indigenous tree species from central India. Biores Technol 68:305–308.

Joutz, F.L. 1992. Biomass fuel supply: A methodology for determining marginal costs. Biores Technol 39:179–183.

Ju, Z., Bramlage, W.J. 1999. Phenolics and lipid-soluble antioxidants in fruit cuticle of apples and their antioxidant activities in model systems. Postharvest Biol Technol 16:107–118.

Kamira, A. 2003. Thermochemical conversion of olive cake to fuel and chemicals. Energy Edu Sci Technol 12:43–48.

Kataki, R., Konwer, D. 2001. Fuelwood characteristics of some indigenous woody species of north-east India. Biomass Bioenergy 20:17–33.

Kawaii, S., Tomono, Y., Katase, E., Ogawa, K., Yano, M. 1999. Antiproliferative effects of the readily extractable fractions prepared from various Citrus juices on several cancer cell lines. J Agric Food Chem 47:2509–2512.

Kim, M.-Y., Choi, S.-W., Chung, S.-K. 2000. Antioxidative flavonoids from the garlic (Allium sativum L.) shoot. Food Sci Biotechnol 9:199–203.

Kim, D.-O., Jeong, S.W., Lee, C.Y. 2003. Antioxidant capacity of phenolic phytochemicals from various cultivars of plums. Food Chem 81:321–326.

Liu, H. 1999. Compacting biomass and municipal solid wastes to form an upgraded fuel. Management Plan for Research Project, Capsule Pipeline Research Centre, University of Missouri, Columbia, MO, pp. 1–43.

Lynd, L.R., van Zyl, W.H., McBride, J.E., Laser, M. 2005. Consolidated bioprocessing of cellulosic biomass: an update. Curr Opin Biotechnol 16:577–583.

McGowan, F. 1991. Controlling the greenhouse effect: the role of renewables. Energy Policy March:111–118.

Moine, C., Gloaguen, V., Gloaguen, J.-M., Granet, R., Krausz, P. 2005. Chemical valorization of forest and agricultural byproducts. Obtention, chemical characteristics, and mechanical behavior of a novel family of hydrophobic films. J Environ Sci Health Part B 39:627–640.

Mohan, D., Pittman, C.U., Steele, P.H. 2006. Pyrolysis of wood/biomass for bio-oil: a critical review. Energy Fuels 20: 848–889.

Moscowitz, C.M. 1978. Source assessment: charcoal manufacturing state of the art, EPA-600/2-78-004z. US Environmental Protection Agency, Cincinnati, OH.

Noureddini, H., Gao, X., Philkana, R.S. 2005. Immobilized Pseudomonas cepacia lipase for biodiesels fuel production from soybean oil. Biores Technol 96:769–777.

Peterson, A.A., Vogel, F., Lachance, R.P., Froling, M., Antal, M.J., Tester, J.W. 2008. Thermochemical biofuel production in hydrothermal media: a review of sub- and supercritical water technologies. Energy Environ Sci 1:32–65.

Shafizadeh, F., Stevenson, T.T. 1982. Saccharification of Douglas-Fir wood by a combination. Of prehydrolysis and pyrolysis. J Appl Polym Sci 27:4577–4485.

Shieh, C.-J., Liao H.-F., Lee, C.-C. 2003. Optimization of lipase-catalyzed biodiesel by response surface methodology. Biores Technol 88:103–106.

Tillman, D.A. 1978. Wood as an energy resource, Academic, New York.

Thomson, A. 2006. An investigation into the implications of using very wet biomass as a fuel. Master's thesis, Mechanical Engineering Department, University of Strathclyde, Glasgow.

Chapter 6
Thermochemical Processes

6.1 Introduction

Biorefineries combine the necessary technologies of biorenewable feedstocks with those of chemical intermediates and final products. Recently there has been renewed interest worldwide in producing biofuels from a range of biorenewable feedstocks.

Char production by pyrolysis, bio-oil production by pyrolysis, gaseous fuels from biomass, Fischer–Tropsch liquids from biomass, hydrothermal liquefaction of biomass, supercritical liquefaction, and biochemical processes of biomass are studied and concluded in this review. Upgraded bio-oil from biomass pyrolysis can be used in vehicle engines as fuel.

The aim of a biorefinery is to extract valuable chemicals and polymers from biomass. The main technologies to produce chemicals from biomass are: (a) biomass refining or pretreatment, (b) thermochemical conversion (gasification, pyrolysis, hydrothermal upgrading), (c) fermentation and bioconversion, and (d) product separation and upgrading.

Biorefineries will use biomass as a feedstock to produce a range of chemicals similar to those currently produced from crude oil in an oil refinery. Over half of the oil produced is used to make transport fuels. In the near future, these are likely to be partly replaced with ethanol or other liquid chemicals produced from sugars and polysaccharides.

There are four main types of biorefineries: biosyngas-based refinery, pyrolysis-based refinery, hydrothermal upgrading-based refinery, and fermentation-based refinery. Biosyngas is a multi-functional intermediate for the production of materials, chemicals, transportation fuels, power and/or heat from biomass.

Thermochemical conversion (TCC) of biomass offers an efficient and economical process to provide gaseous, liquid, and solid fuels and prepare chemicals derived from biomass. TCC technologies have been studied since the eighteenth century with the first patent issued in 1788 by Robert Gardner for his work in the gasification area. However, during the span from 1800–1970 the TCC technologies were forgotten due to an abundance of oil. When the TCC research continued, it began to focus on sources outside of wood and coal (Midgett 2008). TCC technologies include, but are not limited to, gasification, liquefaction, pyrolysis, direct combustion, and supercritical fluid extraction. Gasification and liquefaction continue to be heavily researched and used commercially throughout the world. Researchers are focusing efforts to attempt to understand the complex reaction mechanisms that occur during these processes.

Direct combustion and co-firing with coal for electricity production from biomass is projected to be an important method in the near future. Presently, TCC technologies such as pyrolysis and gasification are not the most important options, since combustion is responsible for over 97% of the world's bioenergy production (Balat 2009). In TCC technologies such as gasification, pyrolysis, hydrothermal upgrading and supercritical water gasification, heat and sometimes pressure are the main drivers of the chemical breakdown of the fuel components. Unlike combustion, in these processes, the fuel is not fully oxidized and the process takes place in the absence of oxygen or in an oxygen-starved environment. Alternatively, the heat to drive the thermal decomposition can be provided indirectly using tubes or other heat exchange mediums.

TCC is characterized by higher temperatures and conversion rates than most other processes. TCC technologies include a continuum of processes ranging from thermal decomposition in a primarily non-reactive environment (commonly called pyrolysis) to decomposition in a chemically reactive environment (usually called gasification if the products are primarily fuel gases). Pyrolysis can be considered an incomplete gasification process, in which a mixture of gaseous, liquid and solid products is produced, each of which may have some immediate use to sustain the process. The characteristics of each of these processes can also vary depending on the oxidizing or reducing media, process temperature and process pressure.

The feedstocks for these systems vary from wood residues, agricultural residues, bark, municipal wastes, pulping liquors to mill effluents. The primary end product of these processes is either a gaseous or liquid fuel that can be cleaned and upgraded to potentially higher-value products or to produce power in a gas turbine or fuel cell. Some TCC processes, such as gasification and pyrolysis, are commercially available while others, such as supercritical gasification, are currently being investigated at the bench scale. However, none of the TCC processes has yet reached its envisioned potential as a commercially successful method to convert biomass to high-value fuels or chemicals.

Thermochemical biomass conversion does include a number of possible roots to produce from the initial biorenewable feedstock useful fuels and chemicals. Thermochemical conversion processes include three subcategories: pyrolysis, gasification, and liquefaction. Figure 6.1 shows the main biomass thermal conversion processes. Biorenewable feedstocks can be used as a solid fuel, or converted into liquid or gaseous forms for the production of electric power, heat, chemicals, or gaseous and liquid fuels. A variety of biomass resources can be converted to liquid, solid and gaseous fuels with the help of some physical, thermochemical, biochemical and biological conversion processes. Main biomass conversion processes are direct liquefaction, indirect liquefaction, physical extraction, thermochemical conver-

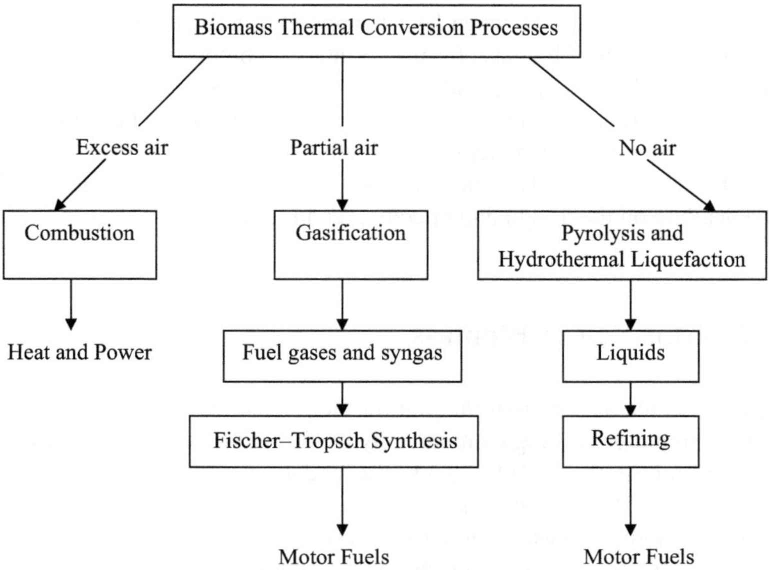

Fig. 6.1 Main biomass thermal conversion processes

sion, biochemical conversion, and electrochemical conversion (Demirbas 2001). Figure 6.2 shows in more detail the biomass thermochemical conversion processes. The conversion of biomass materials has a precise objective to transform a carbonaceous solid material which is originally difficult to handle, bulky and of low energy concentration, into the fuels having physicochemical characteristics which permit economic storage and transferability through pumping systems.

Pyrolysis is the fundamental chemical reaction process that is the precursor of both the gasification and combustion of solid fuels, and is simply defined as the

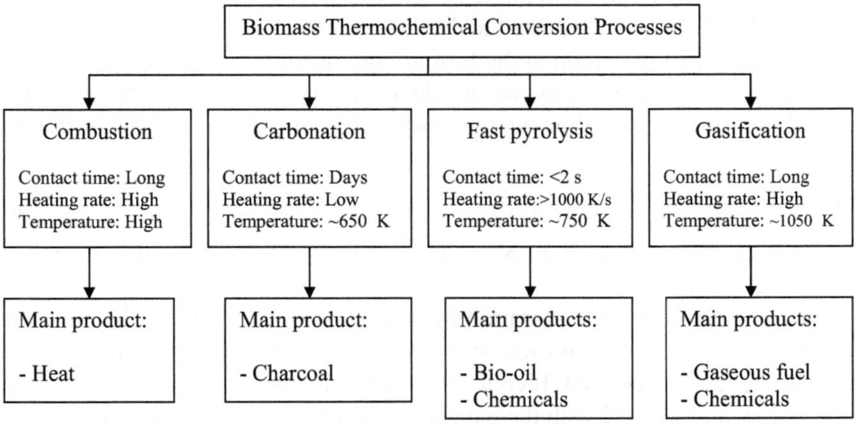

Fig. 6.2 Biomass thermochemical conversion processes with main products

chemical changes occurring when heat is applied to a material in the absence of oxygen. Gasification of biomass for use in internal combustion engines for power generation provides an important alternate renewable energy resource. Gasification is partial combustion of biomass to produce gas and char at the first stage and subsequent reduction of the product gases, chiefly CO_2 and H_2O, by the charcoal into CO and H_2. The process also generates some methane and other higher hydrocarbons depending on the design and operating conditions of the reactor.

6.2 Combustion of Biomass

Combustion is the oxidation of the fuel for the production of heat at elevated temperatures without generating commercially useful intermediate fuel gases, liquids, or solids. Combustion of MSW or other secondary materials is generally referred to as incineration. Particle temperatures in heterogeneous (e.g., unsteady reactions between solid and gas phases) combustion can differ from the surrounding gas temperatures, depending on radiation heat transfer conditions.

Combustion of solids involves the simultaneous processes of heat and mass transport, progressive pyrolysis, gasification, ignition, and burning, with no intermediate steps and with an unsteady, sometimes turbulent, fluid flow. Normally, combustion employs an excess of oxidizer to ensure maximum fuel conversion, but it can also occur under fuel-rich conditions.

Combustion is a basic chemical process that releases energy from a fuel and air mixture. For combustion to occur, fuel, oxygen, and heat must be present together. The combustion is the chemical reaction of a particular substance with oxygen. Combustion represents a chemical reaction, which is a combination of inflammable matter with oxygen of the air, accompanied by heat release. The quantity of heat produced when one mole of a hydrocarbon is burned to carbon dioxide and water is called the heat of combustion. Combustion to carbon dioxide and water is characteristic of organic compounds; under special conditions it is used to determine their carbon and hydrogen content. During combustion the combustible part of the fuel is subdivided into volatile part and solid residue. During heating it evaporates together with a part of carbon in the form of hydrocarbons combustible gases, and carbon monoxide is released by thermal degradation of the fuel. Carbon monoxide is mainly formed by the following reactions: (a) from reduction of CO_2 with unreacted C,

$$CO_2 + C \rightarrow 2CO \tag{6.1}$$

and (b) from degradation of carbonyl fragments (–CO) in the fuel molecules at 600–750 K.

The combustion process is started by heating the fuel above its ignition temperature in the presence of oxygen or air. Under the influence of heat, the chemical bonds of the fuel are cleaved. If complete combustion occurs, the combustible elements (C, H, and S) react with the oxygen content of the air to form CO_2, H_2O and mainly SO_2.

If not enough oxygen is present or the fuel and air mixture is insufficient then the burning gases are partially cooled below the ignition temperature and the combustion process stays incomplete. The flue gases then still contain combustible components, mainly carbon monoxide (CO), unburned carbon (C) and various hydrocarbons (C_xH_y).

The standard measure of the energy content of a fuel is its heating value (HV), sometimes called the calorific value or heat of combustion. In fact, there are multiple values for the HV, depending on whether it measures the enthalpy of combustion (ΔH) or the internal energy of combustion (ΔU), and whether for a fuel containing hydrogen product water is accounted for in the vapor phase or the condensed (liquid) phase. With water in the vapor phase, the lower heating value (LHV) at constant pressure measures the enthalpy change due to combustion. The heating value is obtained by the complete combustion of a unit quantity of solid fuel in an oxygen-bomb calorimeter under carefully defined conditions. The gross heat of combustion or higher heating value (GHC or HHV) is obtained by oxygen-bomb calorimeter method as the latent heat of moisture in the combustion products is recovered.

6.3 Liquefaction Process

In the liquefaction process, biomass is converted to liquefied products through a complex sequence of physical structure and chemical changes. In the liquefaction, biomass is decomposed into small molecules. These small molecules are unstable and reactive, and can repolymerize into oily compounds with a wide range of molecular weight distribution. Liquefaction is a slower form of pyrolysis that is generally performed in a pressurized batch reactor. The product bio-oil from liquefaction tends to be more viscous and has lower oxygen content than the product from fast pyrolysis.

Processes relating to liquefaction of biomass are based on the early research of Appell et al. (1971). These workers reported that a variety of biomass such as agricultural and civic wastes could be converted, partially, into a heavy oil-like product by reaction with water and carbon monoxide/hydrogen in the presence of sodium carbonate.

Liquefaction of biomass is accomplished by natural, direct and indirect thermal, extraction, and fermentation methods. Modern development of the liquefaction process can be traced to the early work at the Bureau of Mines as an extension of coal liquefaction research (Appell et al. 1971). In the case of liquefaction, feedstock macromolecule compounds are decomposed into fragments of light molecules in the presence of a suitable catalyst. At the same time, these fragments, which are unstable and reactive, repolymerize into oily compounds having appropriate molecular weights (Molten et al. 1983).

The pyrolysis and direct liquefaction with water processes are sometimes confused with each other, and a simplified comparison of the two follows. Both are thermochemical processes in which feedstock organic compounds are converted

Table 6.1 Comparison of liquefaction and pyrolysis

Process	Temperature (K)	Pressure (MPa)	Drying
Liquefaction	525–600	5–20	Unnecessary
Pyrolysis	650–800	0.1–0.5	Necessary

into liquid products. In the case of liquefaction, feedstock macromolecule compounds are decomposed into fragments of light molecules in the presence of a suitable catalyst. At the same time, these fragments, which are unstable and reactive, repolymerize into oily compounds having appropriate molecular weights (Molten et al. 1983). With pyrolysis, on the other hand, a catalyst is usually unnecessary, and the light decomposed fragments are converted to oily compounds through homogeneous reactions in the gas phase. The differences in operating conditions for liquefaction and pyrolysis are shown in Table 6.1.

Liquefaction processes result in liquid product, which can be easily stored and transported and require lower process temperatures. Due to these advantages, it is becoming increasingly evident that liquid products offer more potential for the production of bio-based products than gas products and this is reflected in the rapid development of these processes and the large amount of research in this area (Mohan et al. 2006).

6.3.1 Direct Liquefaction

Direct liquefaction of wood by catalyst was carried out in the presence of K_2CO_3 (Ogi et al. 1985). Direct combustion is the oldest method for using biomass, and it still accounts for over 97% of the world's bioenergy production.

Aqueous liquefaction of lignocellulosic materials involves disaggregation of the wood ultrastructure followed by partial depolymerization of the constitutive families (hemicelluloses, cellulose, and lignin). Solubilization of the depolymerized material is then possible (Chornet and Overend 1985). The heavy oil obtained from the liquefaction process was a viscous tarry lump, which sometimes caused troubles in handling. For this purpose, some organic solvents were added to the reaction system. Among the organic solvents tested, propanol, butanol, acetone, methyl ethyl ketone and ethyl acetate were found to be effective on the formation of heavy oil having low viscosity (Demirbas 2000).

The changes during the liquefaction process involve all kinds of processes such as solvolysis, depolymerization, decarboxylation, hydrogenolysis, and hydrogenation. Solvolysis results in micellar-like substructures of the biomass. The depolymerization of biomass leads to smaller molecules. It also leads to new molecular rearrangements through dehydration and decarboxylation. When hydrogen is present, hydrogenolysis and hydrogenation of functional groups, such as hydroxyl groups, carboxyl groups, and keto groups also occur. Figure 6.3 shows the procedures for separation of aqueous liquefaction products.

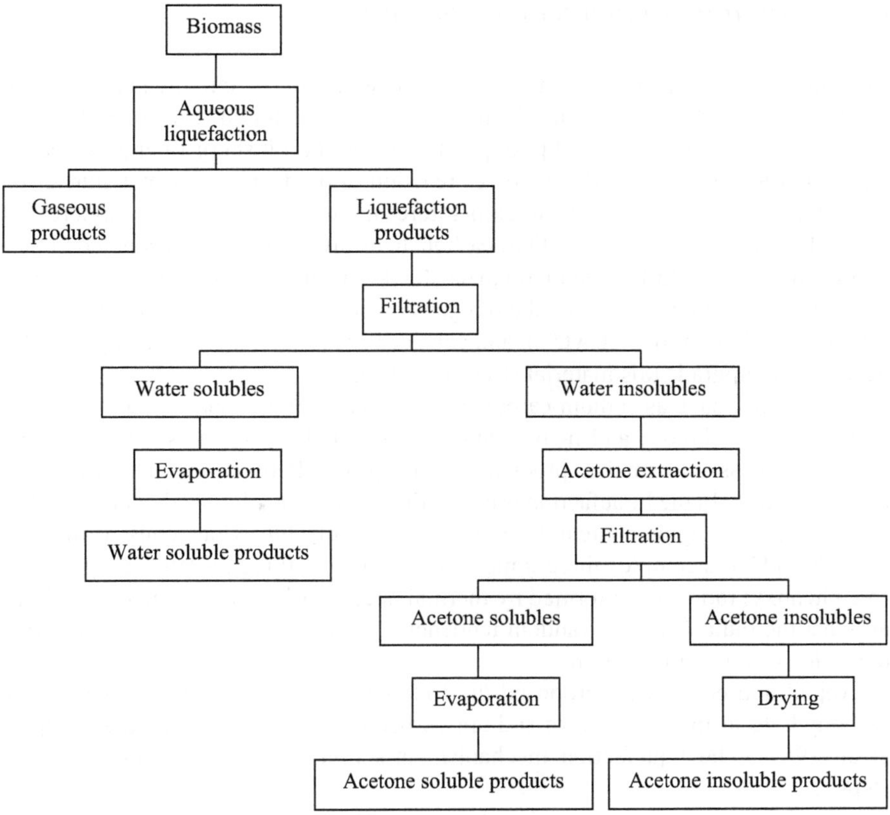

Fig. 6.3 Procedures for separation of aqueous liquefaction products

Research on direct liquefaction has been widely studied in the past, especially in the late 1970s and early 1980s for the purpose of alternative energy production. The feedstocks mainly consisted of wood and municipal solid wastes (MSW). Since then, many aspects of the process are still being studied: the type and condition of various feedstocks, the operating carrier media, and reducing reagents. More specifically researchers are focusing on various operating conditions, such as pH, processing gas, temperature, pressure, catalyst, retention time, solid content, gas-to-volatile solid ratio, and solvents for extraction or processing. In addition to studying these conditions researchers are still focusing efforts to understand the complex reactions that occur during the process. Biomass is complex by nature and varies by location. Developing a process that will handle many biomass sources, and one that is flexible to handle variations of biomass, is desired to increase the potential impact the process may have. Economics currently limit large-scale biomass liquefaction treatment facilities and on-site treatment remains difficult and expensive. However, researchers continue to move forward with their studies and many alternative organic feedstocks have been processed through this technology as a means of waste management as well as renewable energy production.

6.3.2 *Indirect Liquefaction Processes*

Indirect liquefaction involves successive production of an intermediate, such as synthesis gas or ethylene, and its chemical conversion to liquid fuels. In 1983, after several years of laboratory and pilot plant work on the Pittsburg Energy Research Center (PERC) process, which involves reaction of product oil or water slurries of wood particles with H_2 and CO at temperatures up to about 645 K and pressures up to 27 MPa in the presence of sodium carbonate catalyst, researchers concluded that neither process could be commercialized for liquid fuel production without substantial improvement. However, the oxygen content of the resulting complex liquid mixture is still high (6 to 10 wt.%), and considerable processing would appear to be necessary to upgrade this material (Demirbas 1992).

Alkali salts, such as sodium carbonate and potassium carbonate, can initiate the hydrolysis of cellulose and hemicellulose, into smaller fragments. The degradation of biomass into smaller products mainly proceeds by depolymerization and deoxygenation. In the liquefaction process, the amount of solid residue increased in proportion to the lignin content. Lignin is a macromolecule, which consists of alkyl phenols and has a complex three-dimensional structure. It is generally accepted that free phenoxyl radicals are formed by thermal decomposition of lignin above 525 K and that the radicals have a random tendency to form a solid residue through condensation or repolymerization.

Appell and co-workers proposed the following mechanism for liquefaction of carbohydrate in the presence of sodium carbonate and carbon monoxide (Appell et al. 1967). The liquefaction mechanism is summarized in the following four steps:

1. Reaction of sodium carbonate and water with carbon monoxide, to yield sodium formate.
2. Dehydration of vicinal hydroxyl groups in a carbohydrate to an enol, followed by isomerization to ketone.
3. Reduction of newly formed carbonyl group to the corresponding alcohol with formate ion and water.
4. The hydroxyl ion reacts with additional carbon monoxide to regenerate the formate ion.

According to this mechanism, deoxygenation occurs through decarboxylation from ester formed by the hydroxyl group and formate ion derived from the carbonate.

The micellar-like broken down fragments produced by hydrolysis are then degraded to smaller compounds by dehydration, dehydrogenation, deoxygenation and decarboxylation. These compounds, once produced, rearrange through condensation, cyclization and polymerization, leading to new compounds (Chornet and Overend 1985). Russel et al. (1983) recognized the formation of aromatic compounds when cellulose was thermochemically converted in an alkali solution. They suggested that these aromatic compounds were formed by condensation or cyclization of unstable intermediate fragments, which were generated by degradation of the cellulose.

Basically, the alkali process involves reacting carbon monoxide with the lignocellulosic wastes in the presence of a sodium carbonate solution as a catalyst up to a temperature of 644 K and up to 28 MPa. Carbon monoxide reacts with sodium carbonate in the presence of water to form sodium formate, which, in turn, reacts with cellulose in the wood wastes to form oil and regenerate sodium carbonate. The following reactions are believed to be typical:

$$Na_2CO_3 + H_2 + 2CO \rightarrow 2HCOONa + CO_2 \tag{6.2}$$

$$2C_6H_{10}O_5 + 2HCOONa \rightarrow 2C_2H_{10}O_4 + H_2O + CO_2 + Na_2CO_3 \tag{6.3}$$

The liquefaction of biomass has been investigated in the presence of solutions of alkalis (Appell et al. 1971; Demirbas 2000), formate of alkaline metals (Appell et al. 1971; Hsu and Hixon 1981; Kucuk 2005), propanol and butanol (Demirbas 2000; Ogi and Yokokoyama 1993) and glycerol (Hsu and Hixon 1981; Demirbas 1985; Demirbas 1992; Kucuk and Demirbas 1993) or direct liquefaction (Ogi et al. 1985; Minowa et al. 1994; Demirbas 2000). In the case of liquefaction, feedstock macromolecule compounds are decomposed into fragments of light molecules in the presence of a suitable catalyst. At the same time, these fragments, which are unstable and reactive, repolymerize into oily compounds having appropriate molecular weights (Molten et al. 1983).

A variety of gaseous products for use as synthetic fuels and chemicals are produced by gasifying biomass. The actual product composition depends on the biomass composition and the reaction conditions. Catalytic gasification takes advantage of catalysts to serve two primary functions: (a) to increase the yield of gases, at the expense of tar and char, at lower temperatures than are possible without catalysts, and (b) to catalyze secondary reactions to produce the specific product desired. One catalytic approach to producing synthetic fuels and chemicals is indirect liquefaction of biomass, which entails gasifying the biomass to create a synthesis gas consisting of hydrogen and CO_x. These materials, in turn, are converted to the desired liquid fuels and/or chemicals by suitable choice of catalyst, synthesis gas composition, and reaction conditions.

6.3.3 Hydrothermal Conversion Processes

Hydrothermal processing offers a number of potential advantages over other biofuel production methods, including high throughputs, high energy and separation efficiency, the ability to use mixed feedstocks like wastes and lignocellulose, the production of direct replacements for existing fuels, and no need to maintain specialized microbial cultures or enzymes (Peterson et al. 2008).

Hydrothermal technologies are broadly defined as chemical and physical transformations in high-temperature (475–875 K), high-pressure (5–40 MPa) liquid or supercritical water. This thermochemical means of reforming biomass may have energetic advantages, since, when water is heated at high pressures a phase change to steam is avoided which avoids large enthalpic energy penalties. Liquefaction processes are generally lower temperature (475–675 K) reactions, which produce liquid products, often called bio-oil or biocrude. Gasification processes generally take place at higher temperatures (675–975 K) and can produce methane or hydrogen gases in high yields (Peterson et al. 2008).

The hydrothermal upgrading (HTU) process uses water as a medium in which the biomass is heated to approximately 575 K and pressurized to about 4.0 MPa in a batch reactor. The water is condensed to recover heat, and the gases are fired to provide heat for the process. The primary product is a liquid referred to as biocrude. The biocrude itself can be fired in a boiler or a kiln to provide heat, but it can also be upgraded with hydrodeoxygenation. The upgraded product can then be separated into fractions of kerosene (aviation fuel), diesel fuel (automotive fuel) and lubricating oil. Biological chemicals undergo a range of reactions, including dehydration and decarboxylation reactions, which are influenced by the temperature, pressure, concentration, and presence of homogeneous or heterogeneous catalysts. Several biomass hydrothermal conversion processes are in development or demonstration (Peterson et al. 2008).

The HTL or direct liquefaction is a promising technology to treat waste streams from various sources and produce valuable bioproducts such as biocrudes. A major problem with commercializing the HTL processes for biomass conversion today is that it remains uneconomical when compared to the costs of diesel or gasoline production. High transportation costs of large quantities of biomass increase production costs, and poor conversion efficiency coupled with a lack of understanding complex reaction mechanisms inhibits growth of the process commercially.

In the HTU process, biomass is reacted in liquid water at an elevated temperature and pressure. The phase equilibria in the HTU process are very complicated due to the presence of water, supercritical carbon dioxide, alcohols, as well as the so-called biocrude. The biocrude is a mixture with a wide molecular weight distribution and consists of various kinds of molecules. Biocrude contains 10–13% oxygen. The biocrude is upgraded by catalytic hydrodeoxygenation in a central facility. Preliminary process studies on the conversion of various biomass types into liquid fuels have indicated that HTU is more attractive than pyrolysis or gasification. In HTU the biomass, typically in a 25% slurry in water, is treated at temperatures of 575–625 K and 12–18 MPa pressures in the presence of liquid water for 5–20 min to yield a mixture of liquid biocrude, gas (mainly CO_2) and water. Subsequent processing may be able to upgrade the biocrude to useable biofuel. A large proportion of the oxygen is removed as carbon dioxide (Goudriaan and Peferoen 1990).

Biomass, such as wood, with a lower energy density is converted to biocrude with a higher energy density, organic compounds including mainly alcohols and acids, gases mainly including CO_2. Water is also a byproduct. In the products, CO_2, the main component of the gas product, can be used to represent all gas produced, and methanol and ethanol represent organic compounds. In Table 6.2, the weight fraction of each component is assigned on the basis of the data of the vacuum flash of biocrude and the data of a pilot plant (Feng et al. 2004). Figure 6.4 shows the block scheme of a commercial HTU plant. The feedstocks, reaction conditions, and the products for the HTU process are given in Table 6.3.

One of the first HTL studies was conducted by Kranich (1984) using municipal waste materials (MSW) as a source to produce oil. Three different types of materi-

Table 6.2 Representatives for the products from the HTU process

Product	Component	Weight fraction (%)
Biocrude	Polycarbonates	47.5
	Methyl-*n*-propyl ether	2.5
Gas	Carbon dioxide	25.0
Organic compounds	Methanol	5.0
	Ethanol	3.5
Water	Water	16.5

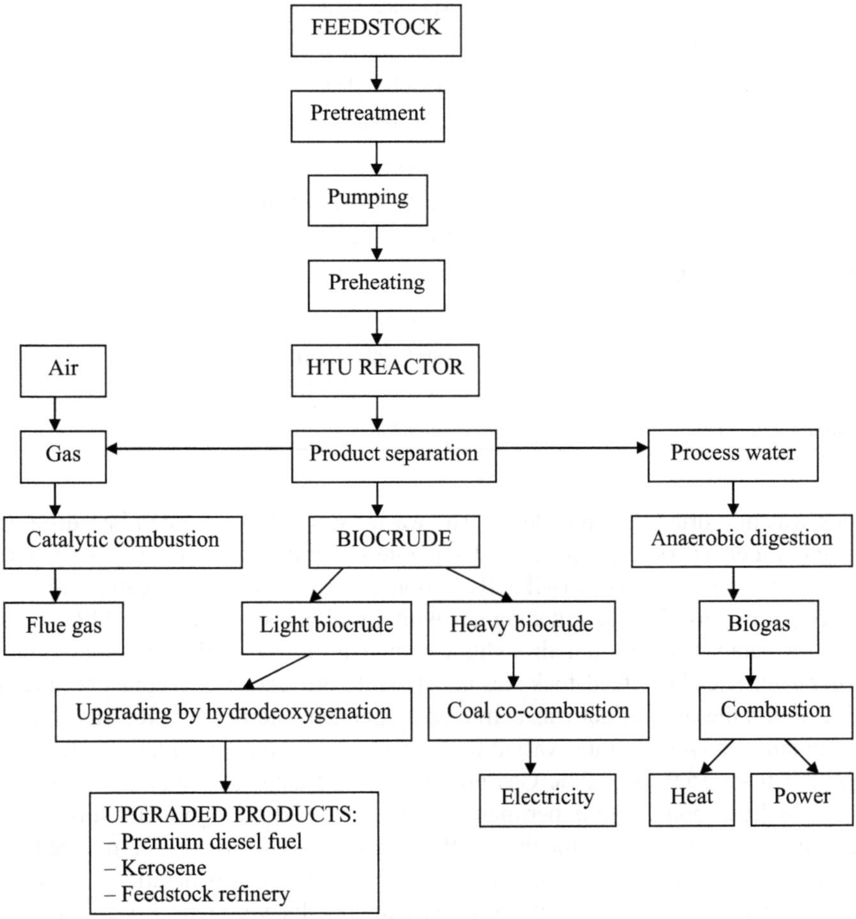

Fig. 6.4 Block scheme of commercial HTU plant

als from a MSW plant were used: primary sewage sludge, settled digester sludge, and digester effluent. Using a magnetically stirred batch autoclave with a hydrogen feed system, slurry feed device, a pressure and temperature recorder, and a wet-test meter for measuring gas product, Kranich processed the waste sources. The feed-

Table 6.3 Feedstocks, reaction conditions, and the products for the HTU process

Biomass feedstocks	Wood and forest wastes
	Agricultural and domestic residues
	Municipal solid wastes
	Organic industrial residues
	Sewage sludge
Reaction conditions	Temperature: 300–350°C
	Pressure: 12–18 MPa
	Resistance time: 5–20 min
	Medium: liquid water
Main chemical reactions	Depolymerization
	Decarboxylation
	Dehydration
	Oxygen removed as CO_2 and H_2O
	Hydrodeoxygenation
	Hydrogenation
Products (wt.% on feedstock)	Biocrude: 45
	Water-soluble organics: 10
	Gas (>90% CO_2): 25
	Process water: 20
Thermal efficiency	70–90%

stock was first dried then powdered. The wastes were also separated into different oil and water slurries and processed separately. Temperatures ranged from 570–720 K with pressures up to 14 MPa. Retention times also varied between 20–90 min. Hydrogen was used as the reducing gas with initial pressures up to 8.3 MPa. Three types of catalyst were studied: sodium carbonate, nickel carbonate, and sodium molybdate. The slurry feedstock was injected into the reactor through a pressurized injector and the oil product was extracted by pentane and toluene. Results showed that organic conversion rates varied from 45 to 99%, and oil production rates were reported from 35.0 to 63.3%. Gas products were found to contain H_2, CO_2, and C_1–C_4 hydrocarbons. The experimental results showed no significant differences between the applications of the three different catalysts. Kranich recommended that the water slurry system was not feasible for scale-up and considerations of a commercial scale process were confined to only the oil slurry system. It was also concluded that no further development work on hydroliquefaction of sewage sludge to oil was necessary. Kranich's recommendation did not hold, mainly due to increases in crude oil prices and the need to find new technologies for energy procurement, and thus many studies on liquefaction of sewage sludge have since been conducted. Research has indicated that liquefaction is a feasible method for the treatment of sewage sludge wastes and has a high oil-producing potential (Suzuki et al. 1986; Itoh et al. 1994; Inoue et al. 1997). Today HTL research is still being conducted with sewage sludge; however, focus has shifted to include many varieties of biomass materials.

Several technologies have been developed to convert biomass into a liquid biofuel with a higher heating value, such as gasification, fast pyrolysis and the HTU. In the HTU, the biomass is treated during 5–20 min with water under subcritical conditions (575–625 K, 10–18 MPa) to give a heavy organic liquid (biocrude) with a heating value of 30–35 MJ/kg. During this process, the oxygen content of the organic material is reduced from about 40% to between 10 and 15%. The removed oxygen ends up in CO_2, H_2O and CO. After 1.6 s at 595 K and 25 MPa, 47% conversion of cellulose in water was obtained yielding hydrolysis products (cellobiose, glucose, and others, 44%) and decomposition products of glucose (erythrose, 1,6-anhydroglucose, 5-hydroxymethylfurfural, 3%). Furthermore, it has been shown that cellobiose decomposes via hydrolysis to glucose and via pyrolysis to glycosylerythrose and glycosylglycolaldehyde, which are further hydrolyzed into glucose, erythrose and glycolaldehyde. Hydrolysis refers to splitting up of the organic particles into smaller organic fragments in water. Hydrothermal decomposition also acts on the large organic molecules reducing them to smaller fragments, some of which dissolve in water.

In the HTU process, biomass chips are pressurized and digested at 475–525 K with recycle water from the process. Subsequently the digested mass is pressurized to 12–18 MPa and reacted in liquid water at 575–675 K for 5–15 min. Under these conditions decarboxylation and depolymerization take place and a biocrude is formed, which separates from the water phase. Part of the process water is recycled. Obviously, the process is very simple with high efficiency.

Hydrothermal reaction involves applying heat under pressure to achieve reaction in an aqueous medium. The treatment of organic wastes by the supercritical water reaction undergoes a homogeneous phase, and interphase mass transfer limitations are avoided and reaction efficiencies of 99.9% can be achieved at residence times lower than 1 min. Because of the distinctive characteristics of water, hydrothermal reaction is an effective method for the treatment of organic wastes. The reaction can be performed under subcritical or supercritical conditions. It can also be classified into two broad categories: (1) oxidative, i. e., involving the use of oxidants, and (2) non-oxidative, i. e., excluding the use of oxidants.

6.3.4 Supercritical Liquefaction

Supercritical fluid extraction (SFE) has potential as an efficacious means of recovering secondary compounds from the lignocellulosic biomass, such as waxy lipids, terpenes and phenolics. Supercritical fluids are less viscous and diffuse more quickly than liquids and can therefore extract plant products more effectively and faster than conventional organic solvents (Gadhe et al. 2007).

Recently, supercritical fluid treatment has been considered to be an attractive alternative in the chemical reaction field. Supercritical water can be realized from the ionic reaction field to the radical reaction field. Therefore, supercritical water can act as a solvent for converting biomass into valuable substances.

Supercritical water (SCW) reforming offers several advantages over the conventional technologies because of the unusual properties of supercritical water. The density of supercritical water is higher than that of steam, which results in a high space-time yield. The higher thermal conductivity and specific heat of supercritical water is beneficial for carrying out the endothermic reforming reactions. In the supercritical region the dielectric constant of water is much lower. Furthermore, the number of hydrogen bonds is much smaller and their strength is considerably weaker. As a result, SCW behaves as an organic solvent and exhibits extraordinary solubility toward organic compounds containing large non-polar groups and most permanent gases. Another advantage of SCW reforming is that the H_2 is produced at a high pressure, which can be stored directly, thus avoiding the large energy expenditures associated with its compression. The supercritical water gasification (SCWG) process becomes economical as the compression work is reduced owing to the low compressibility of liquid feed when compared to that of gaseous H_2. Figure 6.5 shows the schematic setup of the system for supercritical water liquefaction of biomass.

In supercritical water gasification, the reaction generally takes place at the temperature over 875 K and a pressure higher than the critical point of water. With a temperature higher than 875 K, water becomes a strong oxidant, and oxygen in

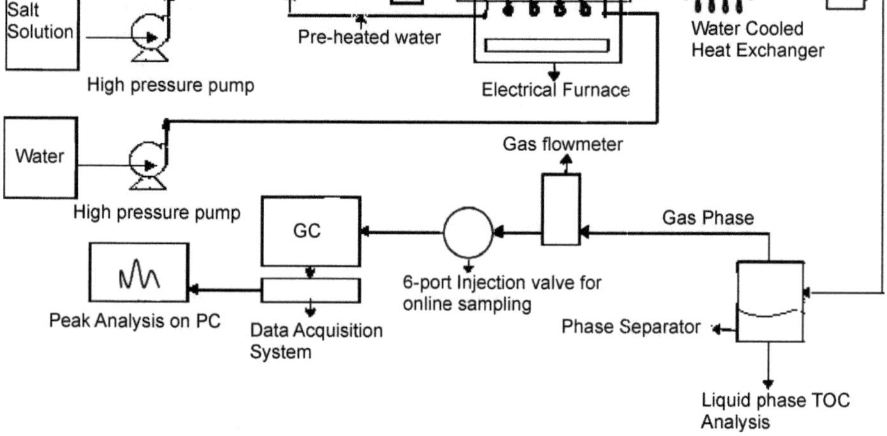

Fig. 6.5 Schematic setup of the system for supercritical water liquefaction of biomass

water can be transferred to the carbon atoms of the biomass. As a result of the high density, carbon is preferentially oxidized into CO_2 but also low concentrations of CO are formed. The hydrogen atoms of water and of the biomass are set free and form H_2. The gas product consists of hydrogen, CO_2, CH_4 and CO.

A general problem with SCWG is the required heat exchange between the reactor outlet and inlet streams. To achieve an acceptable thermal efficiency, it is crucial for the process that the heat of the inlet stream is utilized as far as possible to preheat the feedstock stream (mainly water) to reaction conditions. At the same time, heating of the biomass slurry in the inlet tube of a reactor is likely to cause fouling/plugging problems because the thermal decomposition (>525 K) starts already far below the desired reaction temperature (>875 K) (Byrd et al. 2007). In addition to being a high mass transfer effect, supercritical water also participates in reforming reaction. The molecules in the supercritical fluid have high kinetic energy like the gas and high density like the liquid. Therefore, it is expected that the chemical reactivity can be high. In addition, the ionic product and dielectric constant of supercritical water, which are important parameters for chemical reactions, can be continuously controlled by regulating pressure and temperature. Pressure has a negligible effect on hydrogen yield above the critical pressure of water. As the temperature is increased from 875 to 1075 K the H_2 yield increases from 53 to 73% by volume, respectively. Only a small amount of hydrogen is formed at low temperatures, indicating that direct reformation reaction of ethanol as a model compound in SCW is favored at high temperatures (>975 K). With an increase in the temperature, the hydrogen and carbon dioxide yields increase, while the methane yield decreases. The water excess leads to a preference for the formation of hydrogen and carbon dioxide instead of carbon monoxide. The formed intermediate carbon monoxide reacts with water to hydrogen and carbon dioxide. The low carbon monoxide yield indicates that the water–gas shift reaction approaches completion (Byrd et al. 2007).

The capillaries (1 mm ID and 150-mm-long tubular reactors) are heated rapidly (within 5 s) in a fluidized sand bed to the desired reaction temperature. Experimentation with the batch capillary method has revealed that, especially at low temperatures and high feed concentrations, char formation occurs. A fluidized bed reactor might be a good alternative to solve the problems related to this char and ash formation.

Cellulose and sawdust were gasified in supercritical water to produce hydrogen-rich gas, and Ru/C, Pd/C, CeO_2 particles, nano-CeO_2 and nano-$(CeZr)xO_2$ were selected as catalysts. The experimental results showed that the catalytic activities were Ru/C > Pd/C > nano-$(CeZr)xO_2$ > nano-CeO_2 > CeO_2 particle in turn. The 10 wt.% cellulose or sawdust with CMC can be gasified near completely with Ru/C catalyst to produce 2–4 g hydrogen yield and 11–15 g potential hydrogen yield per 100 g feedstock at the condition of 773 K, 27 MPa, 20 min residence time in supercritical water (Hao et al. 2005).

Catalysts for low-temperature gasification include combinations of stable metals, such as ruthenium or nickel bimetallics and stable supports, such as certain titania, zirconia, or carbon. Without catalyst the gasification is limited (Kruse et al. 2000). Sodium carbonate is effective in increasing the gasification efficiency of cellulose

(Minowa 1997). Likewise, homogeneous alkali catalysts have been employed for high-temperature supercritical water gasification.

Hydrogen is a sustainable, non-polluting source of energy that can be used in mobile and stationary applications. In order to evaluate hydrogen production by SCWG of various types of biomass, extensive experimental investigations have been conducted in recent years. The supercritical water reforming of biomass materials has been found to be an effective technique for the production of hydrogen with short residence times. Hydrogen production by biomass SCW is a promising technology for utilizing high moisture content biomass.

In SWC reforming, the product gases mainly consist of hydrogen and carbon dioxide, with a small amount of methane and carbon monoxide. Hydrogen yields approaching the stoichiometric limit were obtained under catalytic supercritical water conditions. Hydrogen production is influenced by temperature, residence time, and biomass concentration. The high yield of hydrogen was obtained at high reactor temperature, low residence time, and low biomass concentration. The gas product from biomass gasification in catalytic supercritical water contains about 69% H_2 and 30% CO_2 in mole fraction. Others like CH_4 and CO exist in the gas product in smaller amounts. Since a drying process is an energy-intensive operation and the produced H_2 will be stored under high-pressure supercritical water gasification is a promising technology for gasifying biomass with high moisture content.

Figure 6.6 shows the plot for the yield of bio-oils from liquefaction of corn stover at subcritical and supercritical temperatures. The yields of bio-oils increase with increasing temperature. The yield of bio-oils sharply increases between 650 and 680 K and then it reaches a plateau. This temperature range (650–680 K) is in the supercritical zone. At temperatures higher than 690 K the yield of bio-oils are slightly lower. The highest bio-oil yield of 55.7% is obtained at a reaction temperature of 680 K and the yield is 53.3% at 720 K (Demirbas 2008a).

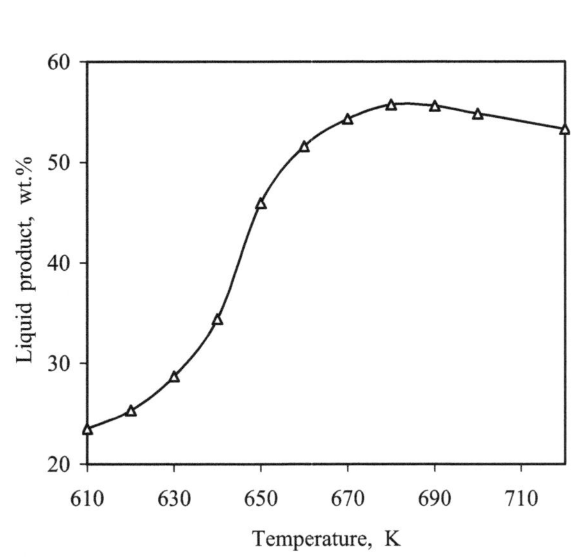

Fig. 6.6 Plot for the yield of liquid products from liquefaction of corn stover at subcritical and supercritical temperatures (liquefaction time of 70 min, water-to-solid ratio of 5:1)

6.4 Pyrolysis Process

Pyrolysis has been used since the dawn of civilization, dating back also to ancient Egyptian times, when tar for caulking boats and certain embalming agents were made by this process. In the 1980s, researchers found that the pyrolysis liquid yield could be increased using fast pyrolysis where a biomass feedstock is heated at a rapid rate and the vapors produced also condensed rapidly (Mohan et al. 2006). If some means is applied to collect the off-gases (smoke), the process is called wood distillation. The ancient Egyptians practiced wood distillation by collecting tars and pyroligneous acid for use in their embalming industry. Pyrolysis of wood to produce charcoal was a major industry in the 1800s, supplying the fuel for the industrial revolution, until it was replaced by coal. In the late nineteenth century and early twentieth century wood distillation was still profitable for producing soluble tar, pitch, creosote oil, chemicals, and non-condensable gases often used to heat boilers at the facility. The wood distillation industry declined in the 1930s due to the advent of the petrochemical industry and its lower-priced products.

Pyrolysis is the thermal decomposition of organic matter occurring in the absence of oxygen or when significantly less oxygen is present than required for complete combustion. This is the basic thermochemical process for converting biomass to a more useful fuel. In the absence of oxygen, biomass is heated or partially combusted in a limited oxygen supply, to produce a hydrocarbon-rich gas mixture, and a carbon-rich solid residue. The products of pyrolysis can be gaseous, liquid, and/or solid. Flash pyrolysis describes the rapid, moderate temperature (675–875 K) pyrolysis that produces liquids. Biomass is heated at rates of 100 to 10 000 K/s and the vapor residence time is normally less than 2 s. The oil products are maximized at the expense of char and gas.

Pyrolysis is a process similar to gasification except generally optimized for the production of fuel liquids (pyrolysis oils) that can be used straight or refined for higher-quality uses such as engine fuels, chemicals, adhesives, and other products. Pyrolysis typically occurs at temperatures in the range of 675 to 975 K. Pyrolysis and combustion of pyrolysis-derived fuel liquids and gases also produce the same categories of end products as direct combustion of solids. Like gasification, their pollution control and conversion efficiencies may be improved.

Pyrolysis and direct liquefaction processes are sometimes confused with each other, and a simplified comparison of the two follows. Both are thermochemical processes in which feedstock organic compounds are converted into liquid products. In the case of liquefaction, feedstock macromolecule compounds are decomposed into fragments of light molecules in the presence of a suitable catalyst. At the same time, these fragments, which are unstable and reactive, repolymerize into oily compounds having appropriate molecular weights (Demirbas 2000). With pyrolysis, on the other hand, a catalyst is usually unnecessary, and the light decomposed fragments are converted to oily compounds through homogeneous reactions in the gas phase.

Fast pyrolysis utilizes biomass to produce a product that is used both as an energy source and a feedstock for chemical production. Considerable efforts have been made to convert wood biomass to liquid fuels and chemicals since the oil crisis in mid-1970s. Most work has been performed on wood, because of its consistency and comparability between tests. However, nearly 100 types of biomass have been tested, ranging from agricultural wastes such as straw, olive pits, and nut shells to energy crops such as miscanthus and sorghum. The pyrolysis reaction rate and the yield of the volatiles are determined by the biomass composition and structure, heating rate, residence time, catalyst, and particle size. Although very fast and very slow pyrolyses of biomass produce markedly different products, the variety of heating rates, temperatures, residence times, and feedstock varieties found in the literature make generalizations difficult to define, in regard to trying to critically analyze the literature (Mohan et al. 2006).

Catalytic cracking is a thermochemical process that employs catalysts using hydrogen-driven reducing reactions to accelerate the breakdown of high molecular weight compounds (e. g., plastics) into smaller products for the purpose of improving selectivity and imparting certain desirable characteristics to the final product, such as volatility and flashpoint of liquid fuels. This cracking process is often employed in oil refinery operations to produce lower molecular weight hydrocarbon fuels from waste feedstocks. These include gasoline from heavier oils, distillation residuals, and waste plastic.

Rapid heating and rapid quenching produce the intermediate pyrolysis liquid products, which condense before further reactions break down higher molecular weight species into gaseous products. High reaction rates minimize char formation. At higher fast pyrolysis temperatures, the major product is gas. If the purpose is to maximize the yield of liquid products resulting from biomass pyrolysis, a low temperature, high heating rate, short gas residence time process would be required. For a high char production, a low temperature, low heating rate process would be chosen. If the purpose were to maximize the yield of fuel gas resulting from pyrolysis, a high temperature, low heating rate, long gas residence time process would be preferred. Table 6.4 shows char, liquid and gaseous products from plant biomass by pyrolysis and gasification. It is believed that as the pyrolysis reaction progresses the carbon residue (semi-char) becomes less reactive and forms stable chemical structures, and consequently the activation energy increases as the conversion level of biomass increases. Pyrolysis liquids are formed by rapidly and simultaneously depolymerizing and fragmenting cellulose, hemicellulose, and lignin with a rapid increase in temperature. Rapid quenching traps many products that would further react (depolymerize, decompose, degrade, cleave, crack or condensate with other molecules) if the residence time at high temperature was extended (Mohan et al. 2006).

Pyrolysis is the simplest and almost certainly the oldest method of processing one fuel in order to produce a better one. Pyrolysis can also be carried out in the presence of a small quantity of oxygen (gasification), water (steam gasification) or hydrogen (hydrogenation). One of the most useful products is methane, which is a suitable fuel for electricity generation using high-efficiency gas turbines.

Cellulose and hemicelluloses form mainly volatile products on heating due to the thermal cleavage of the sugar units. The lignin forms mainly char since it is not readily cleaved to lower molecular weight fragments. The progressive increase in the pyrolysis temperature of the wood led to the release of the volatiles thus form-

Table 6.4 Char, liquid, and gaseous products from plant biomass by pyrolysis and gasification

Thermal degradation	Residence time (s)	Upper temperature (K)	Char	Liquid	Gas
Conventional pyrolysis	1800	470	85–91	7–12	2–5
	1200	500	58–65	17–24	8–14
	900	550	44–49	26–30	16–22
	600	600	36–42	27–31	23–29
	600	650	32–38	28–33	27–34
	600	850	27–33	20–26	36–41
	450	950	25–31	12–17	48–54
Slow pyrolysis	200	600	32–38	28–32	25–29
	180	650	30–35	29–34	27–32
	120	700	29–33	30–35	32–36
	90	750	26–32	27–34	33–37
	60	850	24–30	26–32	35–43
	30	950	22–28	23–29	40–48
Fast pyrolysis	5	650	29–34	46–53	11–15
	5	700	22–27	53–59	12–16
	4	750	17–23	58–64	13–18
	3	800	14–19	65–72	14–20
	2	850	11–17	68–76	15–21
	1	950	9–13	64–71	17–24
Gasification	1800	1250	7–11	4–7	82–89

ing a solid residue that is different chemically from the original starting material (Demirbas 2000). Cellulose and hemicelluloses initially break into compounds of lower molecular weight. This forms an "activated cellulose" which decomposes by two competitive reactions: one forming volatiles (anhydrosugars) and the other char and gases. The thermal degradation of the activated cellulose and hemicelluloses to form volatiles and char can be divided into categories depending on the reaction temperature. Within a fire all these reactions take place concurrently and consecutively. Gaseous emissions are predominantly a product of pyrolitic cracking of the fuel. If flames are present, fire temperatures are high, and more oxygen is available from thermally induced convection.

The biomass pyrolysis is attractive because solid biomass and wastes can be readily converted into liquid products. Products like crude bio-oil or slurry of charcoal of water or oil, have advantages in many aspects of the energy industry such as transport, storage, combustion, retrofitting and flexibility in production and marketing. Among the liquid products, methanol is one of the most valuable products. The point where the cost of producing energy from fossil fuels exceeds the cost of biomass fuels has been reached. With a few exceptions, energy from fossil fuels will cost more money than the same amount of energy supplied through biomass conversion.

Pyrolysis of wood has been studied as a zonal process with zone A (easily degrading zone) occurring at temperatures up to 475 K. The surface of the wood becomes dehydrated at this temperature, and along with water vapor, carbon dioxide, formic acid, acetic acid, and glyoxal are given off. When temperatures of 475 to 535 K are attained, the wood is said to be in zone B and is evolving water vapor, carbon dioxide, formic acid, acetic acid, glyoxal, and some carbon monoxide. The reactions to this point are mostly endothermic, the products are largely non-condensable, and the wood is becoming charred. Pyrolysis actually begins between 535 and 775 K, which is called zone C. The reactions are exothermic, and unless heat is dissipated, the temperature will rise rapidly. Combustible gases such as carbon monoxide from cleaving of carbonyl groups, methane, formaldehyde, formic acid, acetic acid, methanol, and hydrogen are being liberated and charcoal is being formed. The primary products are beginning to react with each other before they can escape the reaction zone. If the temperature continues to rise above 775 K, a layer of charcoal will be formed that is the site of vigorous secondary reactions and is classified as zone D. Carbonization is said to be complete at temperatures of 675 to 875 K. Thermal degradation properties of hemicelluloses, celluloses and lignin can be summarized as follows (Demirbas 2000):

Thermal degradation of hemicelluloses > cellulose >> lignin

If the purpose is to maximize the yield of liquid products resulting from biomass pyrolysis, a low temperature, high heating rate, short gas residence time process would be required. For a high char production, a low temperature, low heating rate process would be chosen. If the purpose is to maximize the yield of fuel gas resulting from pyrolysis, a high temperature, low heating rate, long gas residence time process would be preferred.

6.4.1 Pyrolysis Facilities

Table 6.5 shows the pyrolysis routes and their variants. Conventional pyrolysis is defined as the pyrolysis, which occurs under a slow heating rate. This condition permits the production of solid, liquid, and gaseous pyrolysis products in significant portions. Conventional slow pyrolysis has been applied for thousands of years and has been mainly used for the production of charcoal. The heating rate in conventional pyrolysis is typically much slower than that used in fast pyrolysis. A feedstock can be held at constant temperature or slowly heated. Vapors can be continuously removed as they are formed (Mohan et al. 2006). Slow pyrolysis of biomass is associated with high charcoal content, but the fast pyrolysis is associated with

Table 6.5 Pyrolysis routes and their variants

Method	Residence time	Temperature (K)	Heating rate	Products
Carbonation	Few days	675	Very low	Charcoal
Conventional	5–30 min	875	Low	Char, oil, gas
Slow	20–200	900	High	Oil, char, gas
Fast	0.5–5 s	925	Very high	Bio-oil
Flash liquid[a]	<1 s	<925	High	Bio-oil
Flash gas[b]	<1 s	<925	High	Chemicals, gas
Hydropyrolysis[c]	<10 s	<775	High	Bio-oil
Methanopyrolysis[d]	<10 s	>975	High	Chemicals
Ultrapyrolysis[e]	<0.5 s	1275	Very high	Chemicals, gas
Vacuum pyrolysis	2–30 s	675	Medium	Bio-oil

[a] Flash liquid: liquid obtained from flash pyrolysis accomplished in a time of <1 s
[b] Flash gas: gaseous material obtained from flash pyrolysis within a time of <1 s
[c] Hydropyrolysis: pyrolysis with water
[d] Methanopyrolysis: pyrolysis with methanol
[e] Ultrapyrolysis: pyrolysis with very high degradation rate

tar, at low temperature (675–775 K), and/or gas, at high temperature. At present, the preferred technology is fast or flash pyrolysis at high temperatures with very short residence times (Mohan et al. 2006). Flash pyrolysis utilizes a continuous reactor and can convert 70% of the biomass into bio-oil, while the gaseous and char fractions are each composed of 15% of the original biomass. The gaseous and char fractions are generally used to provide heat for the pyrolysis reactor.

Fast pyrolysis (more accurately defined as thermolysis) is a process in which a material, such as biomass, is rapidly heated to high temperatures in the absence of oxygen. Flash pyrolysis of biomass is the thermochemical process that converts small dried biomass particles into a liquid fuel (bio-oil or biocrude) for almost 75%, and char and non-condensable gases by heating the biomass to 775 K in the absence of oxygen. Char in the vapor phase catalyzes secondary cracking. Table 6.6 shows the range of the main operating parameters for pyrolysis processes.

Table 6.6 Range of the main operating parameters for pyrolysis processes

	Conventional pyrolysis	Fast pyrolysis	Flash pyrolysis
Pyrolysis temperature (K)	550–900	850–1250	1050–1300
Heating rate (K/s)	0.01–1	10–200	>1000
Particle size (mm)	5–50	<1	<0.2
Solid residence time (s)	300–3600	0.5–10	<0.5

Design variables required for fast pyrolysis include the following: feed drying, particle size, pretreatment, reactor configuration, heat supply, heat transfer, heating rates, pyrolysis temperature, vapor residence time, secondary cracking, char separation, ash separation, and liquid collection (Mohan et al. 2006).

Vegetable oils can be converted to liquid and gaseous hydrocarbons by pyrolysis, decarboxylation, deoxygenation, and catalytic cracking processes. Vegetable oils from renewable oilseeds can be used when mixed with diesel fuels, and they can also be used as fuels for diesel engines, but their viscosities are much higher than usual diesel fuel and require modifications of the engines. Different ways have been considered to reduce the viscosity of vegetable oils such as dilution, micro-emulsification, pyrolysis, and transesterification. Compared with transesterification, pyrolysis has more advantages. The liquid fuel produced from pyrolysis has similar chemical components to conventional petroleum diesel fuel. Sunflower seed oil has been pyrolized to obtain liquid fuels at atmospheric pressure, and a reaction temperature of 560–690 K in the presence of zeolite catalysts. The maximum yields of conversion from pyrolysis of sunflower oil were obtained from 2% zeolite catalytic runs at all temperatures. The yield of conversion then decreased with increasing catalyst percentage. The maximum yield of pyrolysis oil was 54% from 2% zeolite catalytic run at 690 K. The liquid fuel produced from pyrolysis has similar chemical components to conventional petroleum diesel fuel (Demirbas 2008b).

6.4.2 Reaction Mechanism of Pyrolysis

Earlier kinetic studies have been conducted under a variety of experimental conditions, resulting in conflicting data with a wide range of kinetic parameters (Stamm 1956; Hofmann and Antal 1984; Desrosiers and Lin 1984; Demirbas 1998; Koullas et al. 1998). Thermogravimetric analysis (TGA) is the general approach applied to determine the weight loss of pyrolyzed samples at various reaction temperatures. Comparison of kinetic data from literature is given in Table 6.7.

The pyrolysis process is always initially endothermic, and almost linear in mass loss. At high heating rates, this character is maintained throughout the process. At low heating rates, the well-known exothermic char-forming processes begin at some point to complete with the basic endothermic nature of the process, and drive it back towards thermoneutrality. There is, however, no evidence of thermodynamically different pathways being followed at high and low heating rates, during the initial stages of pyrolysis. This alone does not assure that a change of mechanism does not occur, but there is also no other evidence to suggest such a change in mechanism with heating rate.

It is believed that as the pyrolysis reaction progresses the carbon residue (semi-char) becomes less reactive and forms stable chemical structures, and consequently the activation energy increases as the conversion level of biomass increases (Tran and Charanjit 1978).

The general changes that occur during pyrolysis are below (Babu and Chaurasia 2003; Mohan et al. 2006):

1. Heat transfer from a heat source, to increase the temperature inside the fuel.
2. The initiation of primary pyrolysis reactions at this higher temperature releases volatiles and forms char.
3. The flow of hot volatiles toward cooler solids results in heat transfer between hot volatiles and cooler unpyrolyzed fuel.
4. Condensation of some of the volatiles in the cooler parts of the fuel, followed by secondary reactions, can produce tar.
5. Autocatalytic secondary pyrolysis reactions proceed while primary pyrolytic reactions simultaneously occur in competition.
6. Further thermal decomposition, reforming, water–gas shift reactions, radicals recombination, and dehydrations can also occur, which are a function of the process's residence time/temperature/pressure profile.

A comparison of pyrolysis, ignition, and combustion of coal and biomass particles reveals the following:

1. Pyrolysis starts earlier for biomass compared with coal.
2. The VM content of biomass is higher compared with that of coal.
3. The fractional heat contribution by VM in biomass is on the order of 70% compared with 36% for coal.
4. Biomass char has more O_2 compared with coal. The fractional heat contribution by biomass is on the order of 30% compared with 70% for coal.
5. The heating value of volatiles is lower for biomass compared with that of coal.
6. Pyrolysis of biomass chars mostly releases CO, CO_2, and H_2O.
7. Biomass has ash that is more alkaline in nature, which may aggravate fouling problems.

The organic compounds from biomass pyrolysis are the following groups:

1. A gas fraction containing CO, CO_2, some hydrocarbons and H_2.
2. A condensable fraction containing H_2O and low molecular weight organic compounds (aldehydes, acids, ketones and alcohols).
3. A tar fraction containing higher molecular weight sugar residues, furan derivatives, phenolic compounds and airborne particles of tar and charred material which form the smoke.

Table 6.7 Comparison of kinetic data from literature

Biomass	Frequency factor (min^{-1})	Activation energy (kJ/mole)	Temp. range (K)	Researcher
Douglas fir sawdust	1.1×10^{11}	105	368–525	(Stamm 1956)
Wood	1.4×10^{6}	84.2	>605	(Barooah and Long 1976)
Douglas fir bark	1.3×10^{10}	100–201	450–850	(Tran and Charanjit 1978)
Missouri sawdust	1.9×10^{7}	95.9	565–665	(Koullas et al. 1998)
Hazelnut shell sawdust	4.7×10^{13}	92–170	450–750	(Demirbas 1998)

The mechanism of pyrolysis reactions of biomass was discussed extensively in an earlier study (Demirbas 2000), where the following were discussed: water is formed by dehydration, methanol arises from the breakdown of methyl esters and/or ethers from decomposition of pectin-like plant materials, methanol also arises from methoxyl groups of uronic acid, and acetic acid is formed in the thermal decomposition of all three main components of wood. When the yield of acetic acid originating from the cellulose, hemicelluloses, and lignin is taken into account, the total is considerably less than the yield from the wood itself. Acetic acid also comes from the elimination of acetyl groups originally linked to the xylose unit.

In the pyrolysis processes, furfural is formed by dehydration of the xylose unit. Quantitatively, 1-hydroxy-2-propanone and 1-hydroxy-2-butanone present high concentrations in the liquid products. These two alcohols are partly esterified by acetic acid. In conventional slow pyrolysis, these two products are not found in so great a quantity because of their low stability. If wood is completely pyrolyzed, resulting products are about what would be expected from pyrolyzing the three major components separately. The hemicelluloses would break down first, at temperatures of 470 to 530 K. Cellulose follows in the temperature range of 510 to 620 K, with lignin being the last component to pyrolyze at temperatures of 550 to 770 K. A wide spectrum of organic substances was contained in the pyrolytic liquid fractions given in the literature (Beaumont 1985). Degradation of xylan yields eight main products: water, methanol, formic, acetic and propionic acids, 1-hydroxy-2-propanone, 1-hydroxy-2-butanone and 2-furfuraldeyde. The methoxy phenol concentration decreased with increasing temperature, while phenols and alkylated phenols increased. The formation of both methoxy phenol and acetic acid was possible as a result of the Diels-Alder cycloaddition of a conjugated diene and unsaturated furanone or butyrolactone.

Timell (1967) described the chemical structure of the xylan as the 4-methyl-3-acetylglucoronoxylan. It has been reported that the first runs in the pyrolysis of the pyroligneous acid consist of about 50% methanol, 18% acetone, 7% esters, 6% aldehydes, 0.5% ethyl alcohol, 18.5% water, and small amounts of furfural (Demirbas 2000).

The composition of the water-soluble products was not ascertained but reported to be composed of hydrolysis and oxidation products of glucose such as acetic acid, acetone, simple alcohols, aldehydes, sugars, etc. (Sasaki et al. 1998). Pyroligneous acids disappear in high-temperature pyrolysis. Levoglucosan is also sensitive to heat and decomposes to acetic acid, acetone, phenols, and water.

6.4.3 Char Production

In reference to wood, pyrolysis processes are alternately referred to as carbonization, wood distillation, or destructive distillation processes. Char or charcoal is produced by slow heating wood (carbonization) in airtight ovens or retorts, in chambers with various gases, or in kilns supplied with limited and controlled amounts of air. As generally accepted, carbonization refers to processes in which the char is the prin-

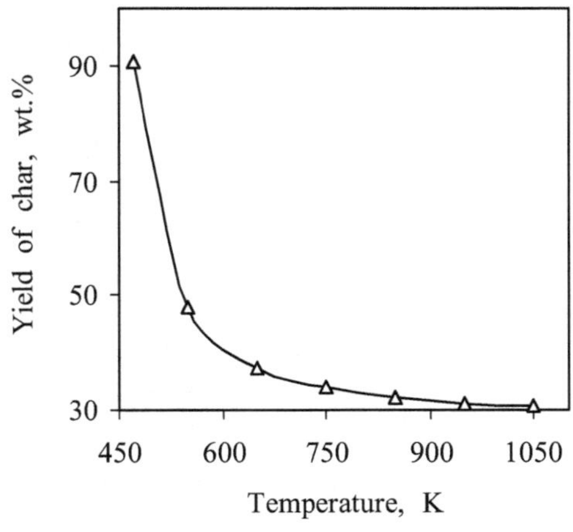

Fig. 6.7 Yield of char from hazelnut shell pyrolysis

cipal product of interest (wood distillation, the liquid; and destructive distillation, both char and liquid). At the usual carbonization temperature of about 675 K, char represents the largest component in wood decomposition products. Typical char from wood contains approximately 80% carbon, 1 to 3% ash, and 12 to 15% volatile components. The char yield decreased gradually from 42.6 to 30.7 % for hazelnut shell and from 35.6 to 22.7% for beech wood with an increase of temperature from 550 to 1150 K, while the char yield from the lignin content decreased sharply from 42.5 to 21.7% until at 850 K during the carbonization procedures. Figure 6.7 shows the yield of char from hazelnut shell pyrolysis (Demirbas 1999). The charcoal yield decreases as the temperature increases. The production of the liquid fraction has a maximum at temperatures between 650 and 750 K. The ignition temperature of charcoal increases as the carbonization temperature increase (Demirbas 2001).

The flow diagram for sustainable char production is presented in Fig. 6.8. Char is manufactured in kilns and retorts. Kilns and retorts generally can be classified as either batch or continuous multiple char manufacturing systems. The continuous multiple systems are more commonly used than are batch systems. Under the sustainable charcoal production, the aim is to minimize material and energy losses at all stages. It is well known that the presence of alkaline cations in biomass affect the mechanism of thermal decomposition during fast pyrolysis causing primarily fragmentation of the monomers making up the natural polymer chains rather than the predominant depolymerization that occurs in their absence.

Wood extractives consist of vegetable oils and valuable chemicals. The vegetable can be converted to biodiesel by transesterification with methanol. The need for increased supplies of charcoal produced from improved and efficient pyrolytic processes is urgent. Recovery of acetic acid and methanol byproducts from pyrolysis was initially responsible for stimulating the charcoal industry. It was shown that hot water washing alone was able to remove a major amount of the alkaline cations (potassium and calcium mainly) from wood (Scott et al. 2000). The wood is then converted into char using improved and efficient kilns after which proper handling is ensured during packaging, storage and transportation to minimize waste. There

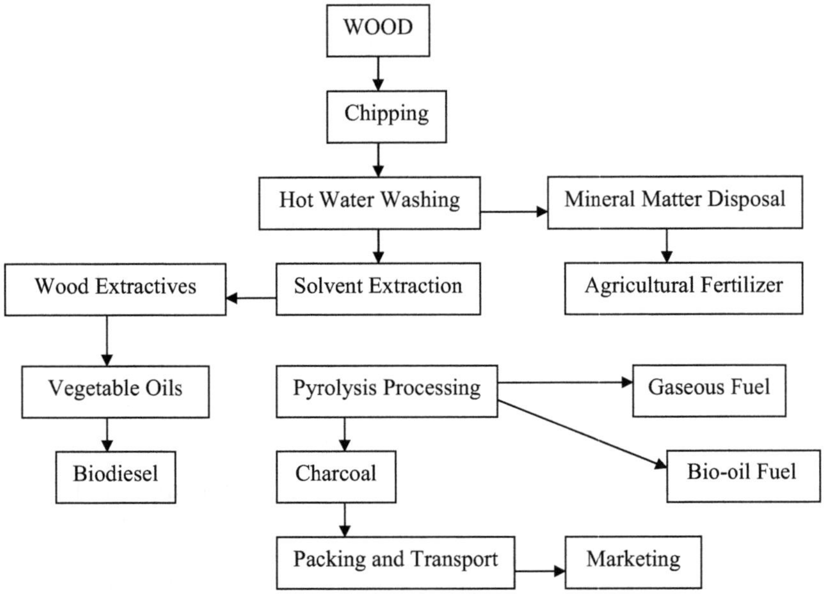

Fig. 6.8 Flow diagram for sustainable char production from wood

are various environmental and socioeconomic benefits associated with each stage in the process.

The elemental composition of char, and its properties, depends on final carbonization temperatures. At increased temperatures, the carbon content increases dramatically. The yield, water absorbency, and hydrogen content decreases rapidly as the carbonization temperature increases. The yield is so low at higher temperatures that the production of charcoal at these temperatures is only of theoretical interest (Demirbas 1999).

Char is very important for developing nations, and for developed ones as well. It is important to learn methods for maximizing its potential, since such a large amount of raw materials are needed for its production. Char is a premium fuel that is widely used in many developing countries to meet household as well as a variety of other needs. It can be readily produced from wood with no capital investment in equipment through the use of charcoal piles, earth kilns, or pit kilns. As the names of these processes suggest, hardwood is carefully stacked in a mound or pit around a central air channel, then covered with dirt, humus, moss, clay, or sod.

6.4.4 Bio-Oil Production

The term bio-oil is used mainly to refer to liquid fuels. Biomass could be converted to liquid, char, and gaseous products via carbonization process at different temperatures. The chemical composition and yields of the char, gas, condensed liquid and tar were determined as function of the carbonization temperature. There are several reasons for bio-oils to be considered as relevant technologies by both developing and industrialized countries: for energy security reasons, environmental concerns, foreign exchange savings, and socioeconomic issues related to the rural sector. Bio-oils are liquid fuels made from biomass materials, such as agricultural crops, municipal wastes and agricultural and forestry byproducts via biochemical or thermochemical processes. The pyrolysis oil (bio-oil) from wood is typically a liquid, almost black through dark red brown. These liquids have a density of about 1200 kg/m^3, which is higher than that of fuel oil and significantly higher than that of the original biomass. Bio-oils also have water contents of typically 15–30 wt.%, which cannot be removed by conventional methods like distillation. Phase separation may occur above certain water contents. The higher heating value (HHV) is below 27 MJ/kg (compared to 43–46 MJ/kg for conventional fuel oils).

Bio-oils are dark brown, free-flowing organic liquids that are comprised of highly oxygenated compounds. Figure 6.9 shows bio-oil produced from flash pyrolysis. The synonyms for bio-oil include pyrolysis oils, pyrolysis liquids, biocrude oil, wood liquids, wood oil, liquid smoke, wood distillates, pyroligneous acid, and liquid wood. Bio-oils contain many reactive species, which contribute to unusual attributes. Chemically, bio-oil is a complex mixture of water, guaiacols, catecols, syringols, vanillins, furancarboxaldehydes, isoeugenol, pyrones, acetic acid, formic acid, and other carboxylic acids. It also contains other major groups of compounds, including hydroxyaldehydes, hydroxyketones, sugars, carboxylic acids, and phenolics (Demirbas 2007).

For example, the bio-oil formed at 725 K contains high concentrations of compounds such as acetic acid, 1-hydroxy-2-butanone, 1-hydroxy-2-propanone, methanol, 2,6-dimethoxyphenol, 4-methyl-2,6-dimetoxyphenol and 2-cyclopenten-1-one, etc. A significant characteristic of the bio-oils was the high percentage of alkylated compounds especially methyl derivatives. As the temperature increased, some of these compounds were transformed via hydrolysis (Kuhlmann et al. 1994). The formation of unsaturated compounds from biomass materials generally involves a variety of reaction pathways such as dehydration, cyclization, Diels-Alder cycloaddition reactions, and ring rearrangement. For example, 2,5-hexandione can undergo cyclization under hydrothermal conditions to produce 3-methyl-2-cyclopenten-1-one with very high selectivity of up to 81% (An et al. 1997).

Fig. 6.9 Bio-oil from flash pyrolysis of biomass

The bio-oil produced by flash pyrolysis is a highly oxygenated mixture of carbonyls, carboxyls, phenolics and water. It is acidic and potentially corrosive. Bio-oil can also be potentially upgraded by hydrodeoxygenation. However, there is a severe energy cost in doing this as a result of its high original oxygen content. Fractionation of bio-oil could potentially result in useful chemical and industrial products. The phenolic fraction of bio-oil, derived from lignin in the biomass, makes up approximately 25% of the organics in the bio-oil. This portion has been identified as a potential source of feedstocks for valuable chemicals such as formaldehyde and resins.

Crude bio-oil can potentially be fired in boilers, kilns, engines and turbines. However, the quality of bio-oil produced is a function of the biomass feed source and its use for these purposes must still be proven for a wide range of feedstocks.

The pyrolysis of biomass is a thermal treatment that results in the production of charcoal, liquid, and gaseous products. Among the liquid products, methanol is one of the most valuable products. The liquid fraction of the pyrolysis products consists of two phases: an aqueous phase containing a wide variety of organo-oxygen compounds of low molecular weight, and a non-aqueous phase containing insoluble organics of high molecular weight. This phase is called tar and is the product of greatest interest. The ratios of acetic acid, methanol, and acetone of the aqueous phase were higher than those of the non-aqueous phase.

Bio-oil has a higher energy density than biomass, and can be obtained by quick heating of dried biomass in a fluidized bed followed by cooling. The byproduct char and gases can be combusted to heat the reactor. Upgraded bio-oil can be used in engines – either totally or partially in a blend.

Biomass is dried and then converted into bio-oil by very quick exposure to heated particles in a fluidized bed. The char and gases produced are combusted to supply heat to the reactor, while the product oils are cooled and condensed. The bio-oil is then shipped by truck from these locations to the hydrogen production facility. It is more economical to produce bio-oil at remote locations and then ship the oil, since the energy density of bio-oil is higher than biomass. For the following analysis, it was assumed that the bio-oil would be produced at several smaller plants, closer to the sources of biomass, such that more affordable feedstocks can be obtained. The liquid yield from hardwood was reduced to 6.2% at 1150 K. Condensed liquid yield of hardwood showed a maximum peak (40.9%) at 850 K and then decreased to 6.2% at higher temperatures. The tar yield from all biomass samples was at about the same level. Table 6.8 shows the characterization of the chemistry and products from pyrolysis and carbonization processes of biomass (Demirbas 2001).

Typical properties and characteristics of wood-derived bio-oil are presented in Table 6.9 (Demirbas 2001). Bio-oil has many special features and characteristics. These require consideration before any application, storage, transport, upgrading or utilization is attempted.

Table 6.10 shows the fuel properties of diesel fuel and biomass pyrolysis oil. The kinematic viscosity of pyrolysis oil varies from as low as 11 cSt to as high as 115 mm^2/s (measured at 313 K) depending on nature of the feedstock, temperature of pyrolysis process, thermal degradation degree, etc. The pyrolysis oils have water contents of typically 15–30 wt.% of the oil mass, which cannot be removed by conventional methods like distillation. Phase separation may partially occur above

Table 6.8 Characterization of chemistry and products of biomass pyrolysis and carbonization

Type	Feature and process	Products and their characterizations
Pyrolysis of holocellulose	General effects: • Color changes from to brown to black • Flexibility and mechanical strength are lost • Size reduced • Weight reduced	Volatile products: • Readily escape during pyrolysis process • 59 compounds produced of which 37 have been identified • CO, CO_2, H_2O, acetal, furfural, aldehydes, ketones
	Processes:	Tar
	Dehydration • Also known as char forming reactions • Produces volatiles products and char	• Levoglucosan is principal component
	Depolymerization • Produces tar Effect of temperature • At low temperatures dehydration predominates • At 630 K depolymerization with production of levoglucosan dominates • Between 550 and 675 K products formed are independent of temperature	Chars • As heating continues there is a 80% loss of weight and remaining cellulose is converted to char • Prolonged heating or exposure to higher temperature (900 K) reduces char formation to 9%
Pyrolysis of lignin	Conventional (carbonization): • At 375–450 K endotherm • At 675 K exotherm • Maximum rate occurring between 625 and 725 K	Char approximately 55%: Distillates (20%) • Methanol – methoxyl groups, acetic acid, acetone Tar (15%) • Phenolic compounds and carboxylic acid Gases • CO, methane, CO_2, ethane
	Fast and flash pyrolysis: • High temperature of 750 K • Rapid heating rate • Finely ground feed material • Less than 10% MC • Rapid cooling and condensation of gases • Yields in 80% range • Char and gas used for fuel	Bio-oil: • Will not mix with hydrocarbon liquids • Cannot be distilled • Substitute for fuel oil and diesel in boilers, furnaces, engines, turbines, etc. Phenol • Utilizes a solvent extraction process to recover phenolics and neutrals • 18–20% of wood weight • Secondary processing of phenol formaldehyde resins • Adhesives • Injected molded plastics

Table 6.8 (*continued*) Characterization of chemistry and products of biomass pyrolysis and carbonization

Type	Feature and process	Products and their characterizations
Pyrolysis of lignin		Other chemicals • Extraction process • Chemical for stabilizing the brightness regression of thermochemical pulp (TMP) when exposed to light • Food flavorings, resins, fertilizers, etc.

Table 6.9 Typical properties and characteristics of wood-derived bio-oil

Property	Characteristics
Appearance	From almost black or dark red-brown to dark green, depending on the initial feedstock and the mode of fast pyrolysis.
Miscibility	Varying quantities of water exist, ranging from ~15 wt.% to an upper limit of ~30–50 wt.% water, depending on production and collection.
	Pyrolysis liquids can tolerate the addition of some water before phase separation occurs.
	Bio-oil cannot be dissolved in water.
	Miscible with polar solvents such as methanol, acetone, etc., but totally immiscible with petroleum-derived fuels.
Density	Bio-oil density is ~1.2 kg/L, compared to ~0.85 kg/L for light fuel oil.
Viscosity	Viscosity of bio-oil varies from as low as 25 cSt to as high as 1000 cSt (measured at 313 K) depending on the feedstock, the water content of the oil, the amount of light ends that have collected, the pyrolysis process used, and the extent to which the oil has been aged.
Distillation	It cannot be completely vaporized after initial condensation from the vapor phase at 373 K or more, it rapidly reacts and eventually produces a solid residue from ~50 wt.% of the original liquid.
	It is chemically unstable, and the instability increases with heating. It is always preferable to store the liquid at or below room temperature; changes do occur at room temperature, but much more slowly and they can be accommodated in a commercial application.
Ageing of pyrolysis liquid	Causes unusual time-dependent behavior.
	Properties such as viscosity increases, volatility decreases, phase separation, and deposition of gums change with time.

certain water contents. The water content of pyrolysis oils contributes to their low energy density, lowers the flame temperature of the oils, leads to ignition difficulties, and, when preheating the oil, can lead to premature evaporation of the oil and resultant injection difficulties. The higher heating value (HHV) of pyrolysis oils is below 26 MJ/kg (compared to 42–45 MJ/kg for conventional petroleum fuel oils). In contrast to petroleum oils, which are non-polar and in which water is insoluble,

Table 6.10 Fuel properties of diesel, biodiesel and biomass pyrolysis oil

Property	Test method	ASTM D975 (diesel)	ASTM D6751 (biodiesel, B100)	Pyrolysis oil (bio-oil)
Flash point	D 93	325 K min	403 K	–
Water and sediment	D 2709	0.05 Max vol.%	0.05 Max vol.%	0.01–0.04
Kinematic viscosity (at 313 K)	D 445	1.3–4.1 mm²/s	1.9–6.0 mm²/s	25–1000
Sulfated ash	D 874	–	0.02 Max wt.%	–
Ash	D 482	0.01 Max wt.%	–	0.05–0.01 wt.%
Sulfur	D 5453	0.05 Max wt.%	–	–
Sulfur	D 2622/129	–	0.05 Max wt.%	0.001–0.02 wt.%
Copper strip corrosion	D 130	No. 3 max	No. 3 max	–
Cetane number	D 613	40 min	47 min	–
Aromaticity	D 1319	–	35 Max vol.%	–
Carbon residue	D 4530	–	0.05 Max mass%	0.001–0.02 wt.%
Carbon residue	D 524	0.35 Max mass%	–	–
Distillation temperature	D 1160	555 K min –611 K max	–	–

biomass oils are highly polar and can readily absorb over 35% water (Demirbas 2007).

Figure 6.10 shows the fractionation of biomass pyrolysis products. The liquid from pyrolysis is often called oil, but is more like tar. This also can be degraded to liquid hydrocarbon fuels. The crude pyrolysis liquid is a thick black tarry fluid. The tar is a viscous black fluid that is a byproduct of the pyrolysis of biomass and is used in pitch, varnishes, cements, preservatives, and medicines as disinfectants and anti-septics. Chemical and physical properties for bio-oils and fuels are given in Table 6.11.

The gas product from pyrolysis usually has a medium heating value (MHV) fuel gas around 15–22 MJ/NM³ or a lower heating value (LHV) fuel gas of around 4–8 MJ/Nm³ from partial gasification depending on feed and processing parameters.

6.4.5 Chemicals and Fuels Production

Table 6.12 shows the gas chromatographic analysis of bio-oil from beech wood pyrolysis (Demirbas 2007). The bio-oil formed at 725 K contained high concentrations of compounds such as acetic acid, 1-hydroxy-2-butanone, 1-hydroxy-2-

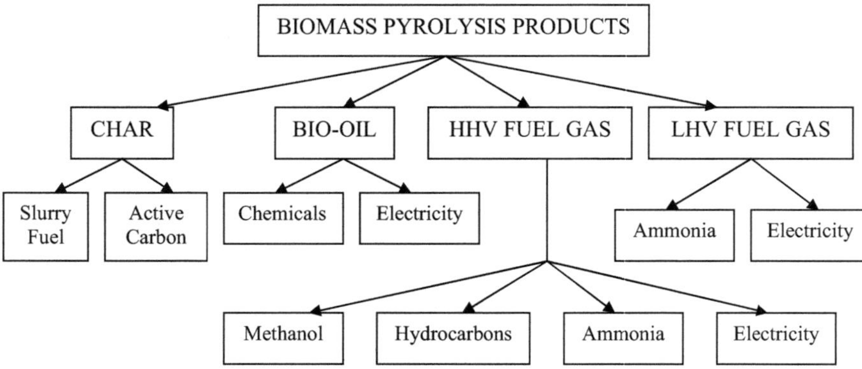

Fig. 6.10 Fractionation of biomass pyrolysis products

Table 6.11 Chemical and physical properties of biomass bio-oils and fuel oils

	Pine	Oak	Poplar	Hardwood	No. 2 oil	No. 6 oil
Elemental (wt.% dry)						
C	56.3	55.6	59.3	58.8	87.3	87.7
H	6.5	5.0	6.6	9.7	12.0	10.3
O (by different)	36.9	39.2	33.8	31.2	0.0	1.2
N	0.3	0.1	0.2	0.2	<0.01	0.5
S	–	0.0	0.00	0.04	0.1	0.8
Proximate (wt.%)						
Ash	0.05	0.05	<0.001	0.07	<0.001	0.07
Water	18.5	16.1	27.2	18.1	0.0	2.3
Volatiles	66.0	69.8	67.4	–	99.8	94.1
Fixed C (by different)	15.5	14.1	5.4	–	0.1	3.6
Viscosity (mm^2/s) at 313 K	44	115	11	58	2.6	–
Density (kg/m^3) at 313 K	1210	1230	–	1170	860	950
Pyrolysis yield (wt.%)						
Dry oil	58.9	55.3	41.0	–	–	–
Water	13.4	10.4	14.6	–	–	–
Char	19.6	12.2	16.4	–	–	–
Gases	13.9	12.4	21.7	–	–	–

propanone, methanol, 2,6-dimethoxyphenol, 4-methyl-2,6-dimetoxyphenol and 2-cyclopenten-1-one, etc. A significant characteristic of the bio-oils was the high percentage of alkylated compounds, especially methyl derivatives.

Table 6.12 Gas chromatographic analysis of bio-oil from beech wood pyrolysis (wt.% dry basis)

Compound	Reaction temperature (K)					
	625	675	725	775	825	875
Acetic acid	16.80	16.5	15.9	12.6	8.42	5.30
Methyl acetate	0.47	0.35	0.21	0.16	0.14	0.11
1-Hydroxy-2-propanone	6.32	6.84	7.26	7.66	8.21	8.46
Methanol	4.16	4.63	5.08	5.34	5.63	5.82
1-Hydroxy-2-butanone	3.40	3.62	3.82	3.88	3.96	4.11
1-Hydroxy-2-propane acetate	1.06	0.97	0.88	0.83	0.78	0.75
Levoglucosan	2.59	2.10	1.62	1.30	1.09	0.38
1-Hydroxy-2-butanone acetate	0.97	0.78	0.62	0.54	0.48	0.45
Formic acid	1.18	1.04	0.84	0.72	0.60	0.48
Guaiacol	0.74	0.78	0.82	0.86	0.89	0.93
Crotonic acid	0.96	0.74	0.62	0.41	0.30	0.18
Butyrolactone	0.74	0.68	0.66	0.67	0.62	0.63
Propionic acid	0.96	0.81	0.60	0.49	0.41	0.34
Acetone	0.62	0.78	0.93	1.08	1.22	1.28
2,3-Butanedione	0.46	0.50	0.56	0.56	0.58	0.61
2,3-Pentanedione	0.34	0.42	0.50	0.53	0.59	0.64
Valeric acid	0.72	0.62	0.55	0.46	0.38	0.30
Iso-valeric acid	0.68	0.59	0.51	0.42	0.35	0.26
Furfural	2.52	2.26	2.09	1.84	1.72	1.58
5-Methyl-furfural	0.65	0.51	0.42	0.44	0.40	0.36
Butyric acid	0.56	0.50	0.46	0.39	0.31	0.23
Iso-butyric acid	0.49	0.44	0.38	0.30	0.25	0.18
Valerolactone	0.51	0.45	0.38	0.32	0.34	0.35
Propanone	0.41	0.35	0.28	0.25	0.26	0.21
2-Butanone	0.18	0.17	0.32	0.38	0.45	0.43
Crotonolactone	0.12	0.19	0.29	0.36	0.40	0.44
Acrylic acid	0.44	0.39	0.33	0.25	0.19	0.15
2-Cyclopenten-1-one	1.48	1.65	1.86	1.96	2.05	2.13
2-Methyl-2-cyclopenten-1-one	0.40	0.31	0.24	0.17	0.13	0.14
2-Methyl-cyclopentenone	0.20	0.18	0.17	0.22	0.25	0.29
Cyclopentenone	0.10	0.14	0.16	0.23	0.27	0.31
Methyl-2-furancarboxaldehyde	0.73	0.65	0.58	0.50	0.44	0.38
Phenol	0.24	0.30	0.36	0.43	0.54	0.66
2,6-Dimethoxyphenol	2.28	2.09	1.98	1.88	1.81	1.76

Table 6.12 (*continued*) Gas chromatographic analysis of bio-oil from beech wood pyrolysis (wt.% dry basis)

Compound	Reaction temperature (K)					
	625	675	725	775	825	875
Dimethyl phenol	0.08	0.13	0.18	0.42	0.64	0.90
Methyl phenol	0.32	0.38	0.44	0.50	0.66	0.87
4-Methyl-2,6-dimetoxyphenol	2.24	2.05	1.84	1.74	1.69	1.58

The influence of temperature on the compounds in liquid products obtained from biomass samples via pyrolysis were examined in relation to the yield and composition of the product bio-oils. The product liquids were analyzed by a gas chromatography-mass spectrometry combined system. The bio-oils were composed of a range of cyclopentanone, methoxyphenol, acetic acid, methanol, acetone, furfural, phenol, formic acid, levoglucosan, guaiacol and their alkylated phenol derivatives. Thermal depolymerization and decomposition of biomass structural components, such as cellulose, hemicelluloses, and lignin form liquids and gas products, as well as a solid residue of charcoal. The structural components of the biomass samples mainly affect the pyrolytic degradation products.

The supercritical water extraction and liquefaction partial reactions also occur during the pyrolysis. Acetic acid is formed in the thermal decomposition of all three main components of biomass. In the pyrolysis reactions of biomass many processes occur: water is formed by dehydration, acetic acid comes from the elimination of acetyl groups originally linked to the xylose unit, furfural is formed by dehydration of the xylose unit, formic acid proceeds from carboxylic groups of uronic acid, and methanol arises from methoxyl groups of uronic acid (Demirbas 2007).

Chemical groups obtained from biomass bio-oil by fast pyrolysis are acids, aldehydes, alcohols, sugars, esters, ketones, phenolics, oxygenates, hydrocarbons, and steroids, shown in Table 6.13.

Table 6.13 Chemicals from biomass bio-oil by fast pyrolysis

Chemical	Minimum (wt.%)	Maximum (wt.%)
Levoglucosan	2.9	30.5
Hydroxyacetaldehyde	2.5	17.5
Acetic acid	6.5	17.0
Formic acid	1.0	9.0
Acetaldehyde	0.5	8.5
Furfuryl alcohol	0.7	5.5
1-Hydroxy-2-propanone	1.5	5.3
Catechol	0.5	5.0

Table 6.13 (*continued*) Chemicals from biomass bio-oil by fast pyrolysis

Chemical	Minimum (wt.%)	Maximum (wt.%)
Methanol	1.2	4.5
Methyl glyoxal	0.6	4.0
Ethanol	0.5	3.5
Cellobiosan	0.4	3.3
1,6-Anhydroglucofuranose	0.7	3.2
Furfural	1.5	3.0
Fructose	0.7	2.9
Glyoxal	0.6	2.8
Formaldehyde	0.4	2.4
4-Methyl-2,6-dimetoxyphenol	0.5	2.3
Phenol	0.2	2.1
Propionic acid	0.3	2.0
Acetone	0.4	2.0
Methylcyclopentene-ol-one	0.3	1.9
Methyl formate	0.2	1.9
Hydroquinone	0.3	1.9
Acetol	0.2	1.7
2-Cyclopenten-1-one	0.3	1.5
Syringaldehyde	0.1	1.5
1-Hydroxy-2-butanone	0.3	1.3
3-Ethylphenol	0.2	1.3
Guaiacol	0.2	1.1

Figure 6.11 shows the vegetable oil upgrading processes. Palm oil has been cracked at atmospheric pressure and a reaction temperature of 723 K to produce bio-oil in a fixed-bed microreactor. The reaction was carried out over microporous HZSM-5 zeolite, mesoporous MCM-41, and composite micromesoporous zeolite as catalysts. The products obtained were gas, liquid (bio-oil), water, and coke. The bio-oil product was composed of hydrocarbons corresponding to gasoline, kerosene, and diesel boiling point range. The maximum conversion of palm oil, 99 wt.%, and gasoline yield of 48 wt.% was obtained with composite micromesoporous zeolite (Sang 2003). The gasoline yield increased with the increase in the Si/Al ratio due to the decrease in the secondary cracking reactions and the drop in the yield of gaseous products. The vegetable oils could be converted to liquid products containing gasoline boiling range hydrocarbons. The results show that the product compositions are affected by catalyst content and temperature.

There are two ethanol production processes that currently employ thermochemical reactions in their processes. The first system is actually a hybrid thermochemi-

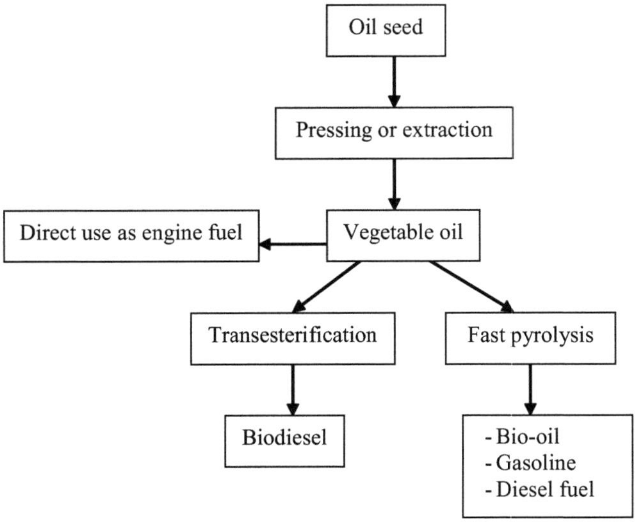

Fig. 6.11 Vegetable oil upgrading processes

cal and biological system. Lignocellulosic biomass materials are first thermochemically gasified and the synthesis gas (a mixture of hydrogen and carbon monoxide) is bubbled through specially designed fermenters. The biomass gasification reaction is:

$$C + H_2O \rightarrow CO + H_2 \tag{6.4}$$

A microorganism that is capable of converting the synthesis gas is introduced into the fermenters under specific process conditions to cause fermentation to bioethanol.

The second thermochemical ethanol production process does not use any microorganisms. In this process, biomass materials are first thermochemically gasified and the synthesis gas passed through a reactor containing catalysts, which cause the gas to be converted into ethanol. Numerous efforts have been made since then to develop commercially viable thermochemical-to-ethanol processes. Ethanol yields up to 50% have been obtained using synthesis gas-to-ethanol processes. Some processes that first produce methanol and then use catalytic shifts to produce ethanol have obtained ethanol yields in the range of 80%. Unfortunately, like the other processes, finding a cost-effective all-thermochemical process has been difficult.

6.4.6 Upgrading of Pyrolysis Products

The overview of a fast pyrolysis liquids-based biorefinery is given in Fig. 6.12. Biorefineries for the purpose of producing several products and byproducts such as biofuels, heat and/or electricity, have been in focus in the recent years. In a biore-

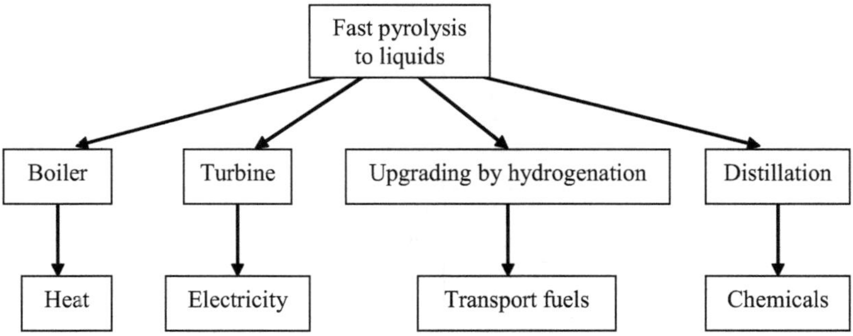

Fig. 6.12 Overview of fast pyrolysis liquids-based biorefinery

finery, biomass can be converted to useful biomaterials and/or energy carriers in an integrated manner and thereby maximize the economic value of the biomass used while reducing the waste streams produced (Zhang 2008).

Figure 6.13 shows biomass upgrading by pyrolysis. Upgrading of condense liquid from biomass includes three stages: physical upgrading (differential condensation, liquid filtration and solvent addition), catalytic upgrading (deoxygenating and reforming), and chemical upgrading (new fuel and chemicals synthesis).

The bio-oil obtained from the fast pyrolysis of biomass has a high oxygen content. Ketones and aldehydes, carboxylic acids and esters, aliphatic and aromatic alcohols, and ethers have been detected in significant quantities. Because of the reactivity of oxygenated groups, the main problems of the oil are instability. Therefore study of the deoxygenation of bio-oil is needed. In the previous work the mechanism of hydrodeoxygenation (HDO) of bio-oil in the presence of a cobalt molybdate catalyst was studied (Zhang et al. 2003).

The main hydrodeoxygenation (HDO) reaction is represented in Eq. 6.5:

$$-(CH_2O)- + H_2 \rightarrow -(CH_2)- + H_2O \tag{6.5}$$

This is the most important route of chemical upgrading. The reaction in Eq. 6.5 is analogous to typical refinery hydrogenations like hydrodesulfurization and hydrodenitrification. In general, most of the hydrodeoxygenation studies have been performed using existing hydrodesulfurization catalysts (NiMo and CoMo on suitable carriers). Such catalysts need activation using a suitable sulfur source and this is a major drawback when using nearly sulfur-free resources like bio-oil. Table 6.14 shows the upgrading processes of flash pyrolysis liquid or bio-oil.

Hydroprocessing is the term used for the catalytic reactions of hydrogen with process streams to remove the heteroatoms sulfur, nitrogen, oxygen, and metals, and for the saturation of aromatic structures. Hydroprocessing can also refer to hydrogenation of olefins or other unsaturated species. Hydroprocessing is carried out in fixed-bed reactors at high pressures, with severe reaction conditions generally increasing with the heaviness of the feed. Reactors used with the heavier feeds are trickle beds, in which the liquid oil trickles through the catalyst bed, with hydrogen concentrated in the gas phase. Slurry reactors have also been used for heavy feedstocks.

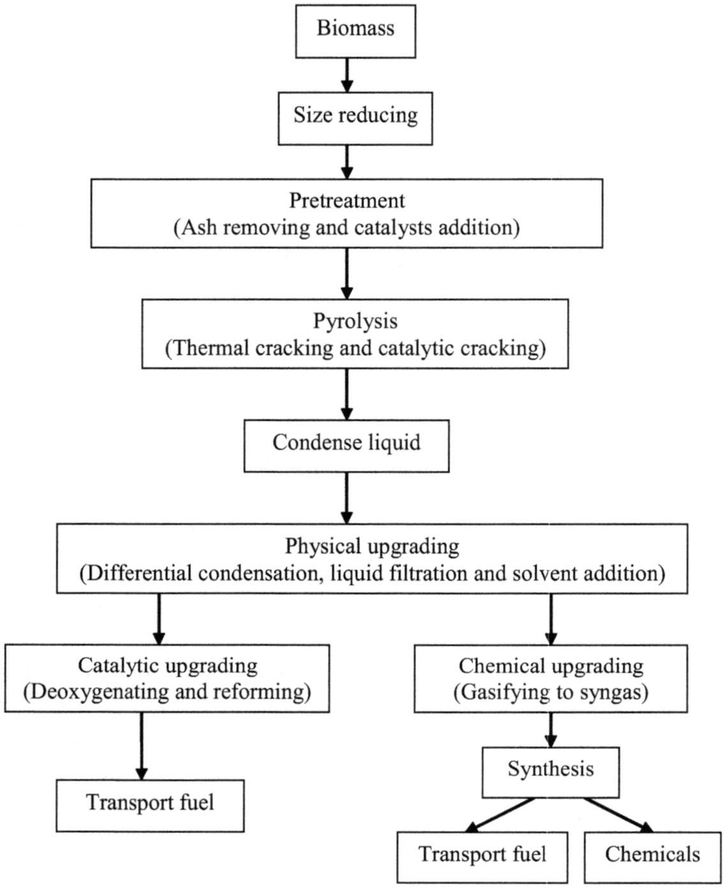

Fig. 6.13 Biomass upgrading by pyrolysis

Table 6.14 Upgrading processes of flash pyrolysis liquid or bio-oil

Upgrading process	Products
Hydrodeoxygenation (HDO)	Fuel, chemicals (phenolics)
Homogenous catalyst hydrogenation	Stabilized oil
Alcoholysis and reactive distillation	Fuel
Reactive extraction	Chemicals (organic acids)

The primary objective, catalytic partial hydrodeoxygenation, is to increase the energetic value of the oil by removing bound oxygen in the form of water. Hydrodeoxygenation of bio-oils involves treating the oils at moderate temperatures with high-pressure hydrogen in the presence of heterogeneous catalysts. The process was carried out in two distinct stages, a first stage at relatively low temperatures (525–575 K), aimed to stabilize the bio-oil and a second stage at higher tempera-

Table 6.15 Properties of the raw and upgraded bio-oil

Property	Raw bio-oil	Upgraded bio-oil
Density (kg/L)	1.12	0.93
Elemental analysis (wt.%)		
C	60.4	87.7
H	6.9	8.9
O	41.8	3.0
N	0.9	0.4
Higher heating value (MJ/kg)	21.3	41.4

tures (575–675 K) to deoxygenate the intermediate product. Different types of catalysts were screened, ranging from conventional sulfided catalysts used in the HDS process (i. e., $NiMo/Al_2O_3$, $CoMo/Al_2O_3$) to novel non-sulfided catalysts based on noble metal catalysts (i. e., Ru/Al_2O_3). Operating conditions were optimized to obtain the highest yield of a hydrocarbon like liquid product.

Hydrogen can be used to convert biomass to fuels, which requires a higher level of hydrodeoxygenation compared to the conversion of fossil feedstocks to fuels. "Green diesel" refers to an acceptable diesel pool blend component produced from a suitable biofeedstock. Green diesel can be produced by either hydrodeoxygenation or decarboxylation of plant oils and greases with propane as a co-product. Although hydrodeoxygenation removes oxygen from a triglyceride or free fatty acid by reaction with hydrogen to form water and an *n*-paraffin, decarboxylation removes oxygen in the biomass carboxy groups to form carbon dioxide and a shorter *n*-paraffin (Kaparaju et al. 2009).

Properties of the raw and upgraded bio-oil are given in Table 6.15. As seen in this table, properties of the raw and upgraded bio-oil are highly different.

6.4.7 Refining of Pyrolysis Products

Various aspects of biorefinery research and development since the beginning of the 1990s are presented. Many of the currently used biorenewable-based industry products are results of direct physical, catalytical or chemical treatment and processing of biomass feedstocks.

The primary objective is the refining of pyrolysis products to obtain valuable fuels like gasoline, diesel fuel and jet fuel, and chemicals from biomass. Figure 6.14 shows the products from biomass by pyrolysis-based refinery.

In vacuum pyrolysis, biomass is heated in a vacuum in order to decrease the boiling point and avoid adverse chemical reactions. In flash vacuum thermolysis (FVT), the residence time of the substrate at the working temperature is limited as much as possible, again in order to minimize secondary reactions. Figure 6.15 shows the multiple stages vacuum pyrolysis of biomass.

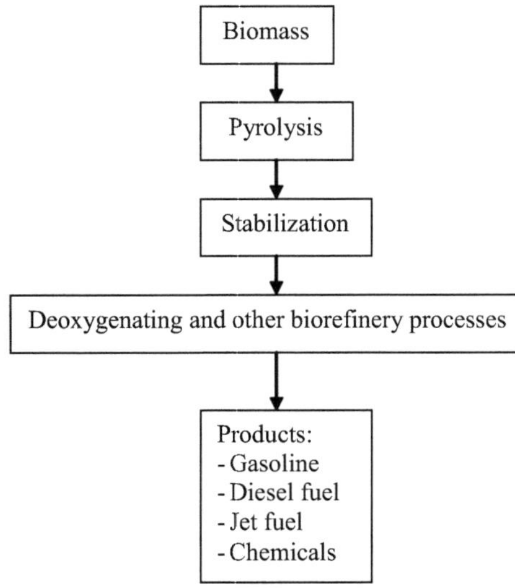

Fig. 6.14 Products from biomass by pyrolysis-based refinery

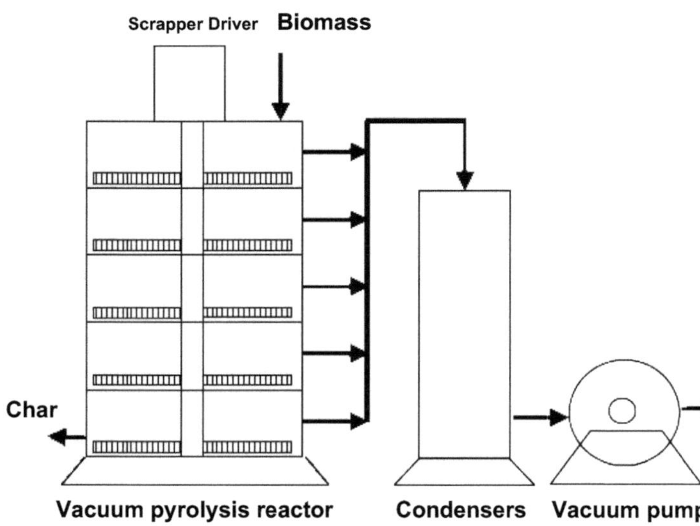

Fig. 6.15 Multiple stages vacuum pyrolysis of biomass

Vacuum pyrolysis is the thermal degradation of a feedstock in the absence of oxygen and under low pressure, to produce a bio-oil and char as main products, together with water and non-condensable gases. Both the bio-oil and char have a high-energy content and may be used as fuels. An incredible number of chemical compounds are also found in the bio-oil and these compounds can be extracted and sold as high-value chemicals.

Vacuum pyrolysis is a relatively new variant of pyrolysis with many recycling applications. During vacuum pyrolysis of biomass, the biorenewable feedstock is thermally decomposed under reduced pressure. The quick removal of the vapors reduces the residence times of the macromolecules and hence minimizes secondary

Table 6.16 Comparison of pyrolysis and gasification processes

		Products (wt.%)		
Process	Conditions	Liquid	Char	Gas
Fast pyrolysis	Moderate temperature (700–800 K)	70–77	10–14	11–15
	Short hot vapor residence time (<2 s)			
Slow pyrolysis	Low-moderate temperature (600–725 K)	27–33	33–38	33–38
	Long residence time			
Vacuum pyrolysis	Moderate temperature (675 and 775 K)	74–80	8–12	9–14
	Low pressure (<0.15 atm)			
Gasification	High temperature (>1075 K)	4–6	8–12	82–88
	Long residence time			

decomposition reactions such as cracking, repolymerization and recondensation, which occur during atmospheric pyrolysis. Temperatures between 675 and 775 K and pressures of about 0.15 atmospheres are typically used. Table 6.16 gives a comparison of pyrolysis and gasification processes.

6.4.8 Refining Opportunities for Bio-Oil

The bio-oil contains the thermally cracked products of the original cellulose, hemicelluloses, and lignin fractions present in the biomass. It also contains a high percentage of water, often as high as 30%. The total oil is often homogeneous after quenching but can easily be separated into two fractions: a water-soluble fraction and a heavier pyrolytic lignin fraction. The addition of more water allows the pyrolytic lignin fraction to be isolated and the majority of it consists of the same phenolic polymer as lignin but with smaller molecular weight fragments. Pyrolytic lignin is a better feedstock for liquid fuel production than the water-soluble fraction because of its lower oxygen content and therefore the study focused on evaluating it as a potential feedstock for the production of highly aromatic gasoline.

6.5 Gasification Process

Gasification is a form of pyrolysis, carried out at high temperatures in order to optimize the gas production. The resulting gas, known as producer gas, is a mixture of carbon monoxide, hydrogen and methane, together with carbon dioxide and nitrogen. Gasification of biomass is a well-known technology that can be classified depending on the gasifying agent: air, steam, steam-oxygen, air-steam, O_2-enriched air, etc.

Biomass gasification technologies have historically been based upon partial oxidation or partial combustion principles, resulting in the production of a hot, dirty,

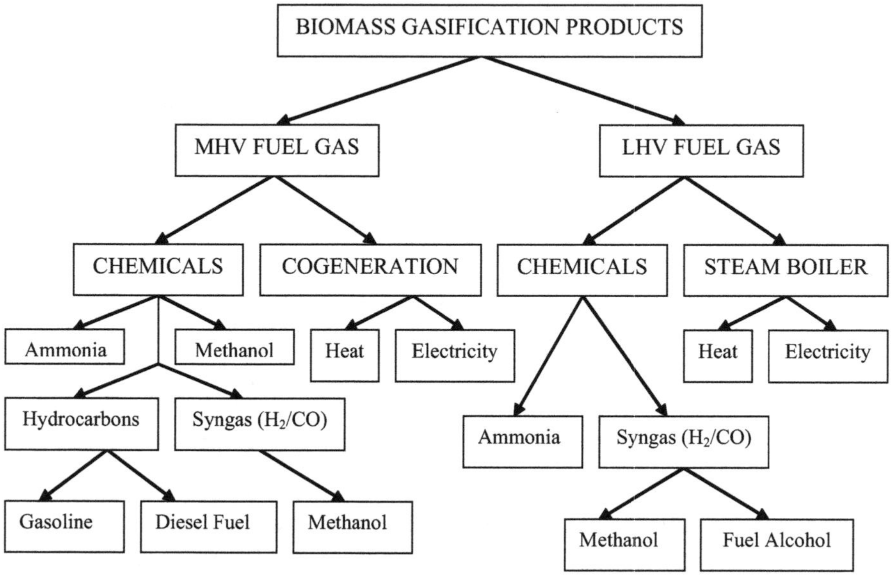

Fig. 6.16 Fractionation of biomass gasification products

low heating value gas that must be directly ducted into boilers or dryers. In addition to limiting applications and often compounding environmental problems, these technologies are an inefficient source of usable energy.

Most biomass gasification systems utilize air or oxygen in partial oxidation or combustion processes. These processes suffer from low thermal efficiencies and low calorific gas because of the energy required to evaporate the moisture typically inherent in the biomass and the oxidation of a portion of the feedstock to produce this energy. In gasification, the aim is to convert biomass entirely into the gas phase. Either the fuel is partially oxidized directly to produce the heat, or heat is provided indirectly in conjunction with direct steam injection to drive its decomposition to gaseous fuel products such as carbon monoxide, hydrogen, methane and some higher weight hydrocarbons, collectively referred to as tars. If air or oxygen is used directly in the gasifier, generally only about 30% of the stoichiometric oxygen required for full oxidation is supplied to a gasifier. As a result of this oxidation, the gaseous product also contains the non-fuel components carbon dioxide, water and, if air is used, nitrogen. Potential biomass fuel sources for gasification are quite abundant. One of the great advantages of gasification is its ability to process rather diverse and heterogeneous feedstocks. Moisture content is the most critical raw fuel parameter to ensure a higher-quality product gas.

Figure 6.16 shows the fractionation of biomass gasification products. The biomass fuels are suitable for the highly efficient power generation cycles based on gasification and pyrolysis processes. A product gas from thermal decomposition is composed of CO, CO_2, H_2O, H_2, CH_4, other gaseous hydrocarbons (CHs), tars, char, inorganic constituents, and ash. Gas composition of product from the biomass gasification depends heavily on the gasification process, the gasifying agent, and the feedstock composition. The relative amount of CO, CO_2, H_2O, H_2, and (CHs) depends on the stoichiometry of the gasification process. If air is used as the gasifying agent, then roughly half of the product gas is N_2. The air/fuel ratio in a gasification process generally ranges from 0.2 to 0.35 and if steam is the gasifying agent, the steam/biomass ratio is around 1. The actual amount of CO, CO_2, H_2O, H_2, tars, and (CHs) depends on the partial oxidation of the volatile products, as shown in Eq. 6.6:

$$C_nH_m + (n/2+m/4)\,O_2 \leftrightarrows nCO + (m/2)\,H_2O \tag{6.6}$$

The major thermochemical reactions include the following:

Steam and methane:

$$CH_4 + H_2O \leftrightarrows CO + 3H_2 \tag{6.7}$$

Water–gas shift:

$$CO + H_2O \leftrightarrows CO_2 + H_2 \tag{6.8}$$

Carbon char to methane:

$$C + 2H_2 \leftrightarrows CH_4 \tag{6.9}$$

Carbon char oxides (Boudouard reaction):

$$C + CO_2 \leftrightarrows 2CO \tag{6.10}$$

6.5.1 Gasification Facilities

The relative simplicity of the gasification system enables its operation to be within the technical expertise of most operators who are experienced with conventional boilers and furnaces, and results in favorable project economics. Its modular design allows a wide range of scale-up or scale-down possibilities, so the systems can vary in size from about one ton per hour of residue to 20 tons per hour or larger, with the size being limited only by biomass availability.

The main steps in the gasification process are:

- Step 1. Biomass is delivered to a metering bin from which it is conveyed with recycled syngas or steam, without air or oxygen into the gasifier.
- Step 2. The material is reformed into a hot syngas that contains the inorganic (ash) fraction of the biomass and a small amount of unreformed carbon.
- Step 3. The sensible heat in the hot syngas is recovered to produce heat for the reforming process.
- Step 4. The cool syngas passes through a filter and the particulate in the syngas is removed as a dry, innocuous waste. The clean syngas is then available for combustion in engines, turbines, or standard natural gas burners with minor modifications.

For electricity generation, the two most competitive technologies are direct combustion and gasification. Typical plant sizes at present range from 0.1 to 50 MW. Co-generation applications are very efficient and economical. Fluidized bed com-

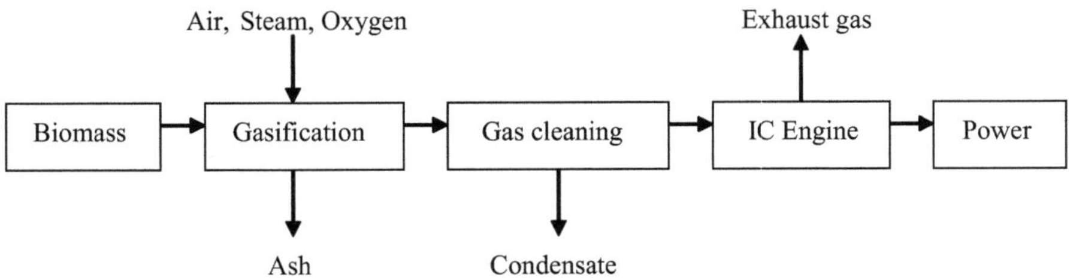

Fig. 6.17 System for power production by means of biomass gasification

bustion (FBC) is efficient and flexible in accepting varied types of fuels. Gasifiers first convert solid biomass into gaseous fuels, which is then used through a steam cycle or directly through gas turbine/engine. Gas turbines are commercially available in sizes ranging from 20 to 50 MW (Badin and Kirschner 1998).

Commercial gasifiers are available in a range of size and types, and run on a variety of fuels, including wood, charcoal, coconut shells and rice husks. Power output is determined by the economic supply of biomass, which is limited to 80 MW in most regions (Overend 1998). Figure 6.17 shows the system for power production by means of biomass gasification. The gasification system of biomass in fixed-bed reactors provides the possibility of combined heat and power production in the power range of 100 kWe up to 5 MWe. A system for power production by means of fixed-bed gasification of biomass consists of the main unit gasifier, the gas cleaning system, and engine.

Gasifiers are used to convert biomass into a combustible gas. The combustible gas is then used to drive a high efficiency, combined-cycle gas turbine. Combustible gas energy conversion devices that are discussed are reciprocating engines, turbines, microturbines, fuel cells, and anaerobic digesters. The resulting combustible gas can be burnt to provide energy for cooking and space heating, or create electricity to power other equipment. Since many of the parasites and disease-producing organisms in the waste are killed by the relatively high temperature in the digester tanks, the digested material can also be used as fertilizer or fish feed.

Various gasification technologies include gasifiers where the biomass is introduced at the top of the reactor and the gasifying medium is either directed co-currently (downdraft) or counter-currently up through the packed bed (updraft). Other gasifier designs incorporate circulating or bubbling fluidized beds. Table 6.17 shows the fixed-bed and fluidized-bed gasifiers and reactor types used in gasification pro-

Table 6.17 Fixed-bed and fluidized-bed gasifiers and reactor types using in gasification processes

Gasifier	Reactor type
Fixed bed	Downdraft, updraft, co-current, counter-current, cross-current, others
Fluidized bed	Single reactor, fast fluid bed, circulating bed

Table 6.18 Composition of gaseous products from various biomass fuels by different gasification methods (% by volume)

H_2	CO_2	O_2	CH_4	CO	N_2
10–19	10–15	0.4–1.5	1–7	15–30	43–60

cesses. Table 6.18 shows composition of gaseous products from various biomass fuels by different gasification methods.

Numerous types of gasifiers have been developed and tested and many industrial applications can use the technology. Gasifiers have been built and operated using a wide variety of configurations including:

1. Fixed-bed (updraft or downdraft fixed beds) gasifiers
2. Fluidized-bed (fluidized or entrained solids serve as the bed material) gasifiers
3. Others including moving grate beds and molten salt reactors

Schematic diagrams of updraft (counter-current) and downdraft (co-current) fixed-bed gasifiers are shown in Figs. 6.18 and 6.19, respectively. Other gasifier designs incorporate circulating or bubbling fluidized beds.

In the drying zone, feedstock descends into the gasifier and moisture is removed using the heat generated in the zones below by evaporation. In the distillation zone, pyrolysis and partial oxidation takes place using the thermal energy released by the partial oxidation of the pyrolysis products. Tar yields can range from 0.1% (down-

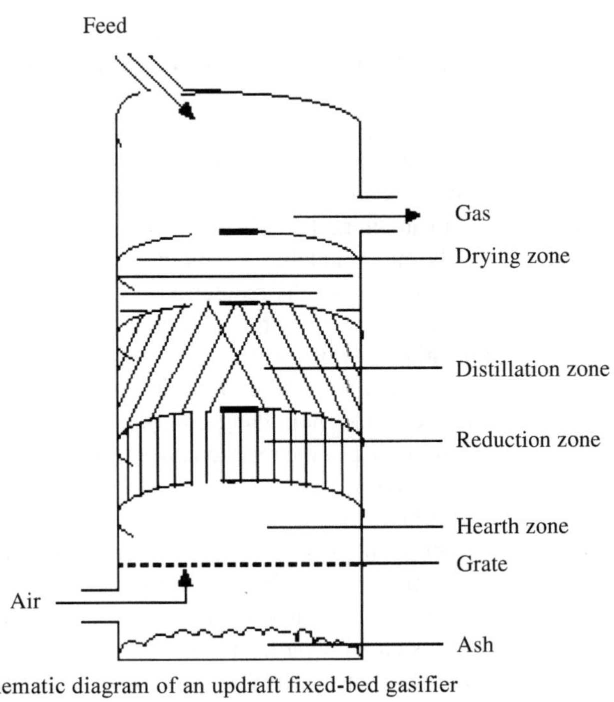

Fig. 6.18 Schematic diagram of an updraft fixed-bed gasifier

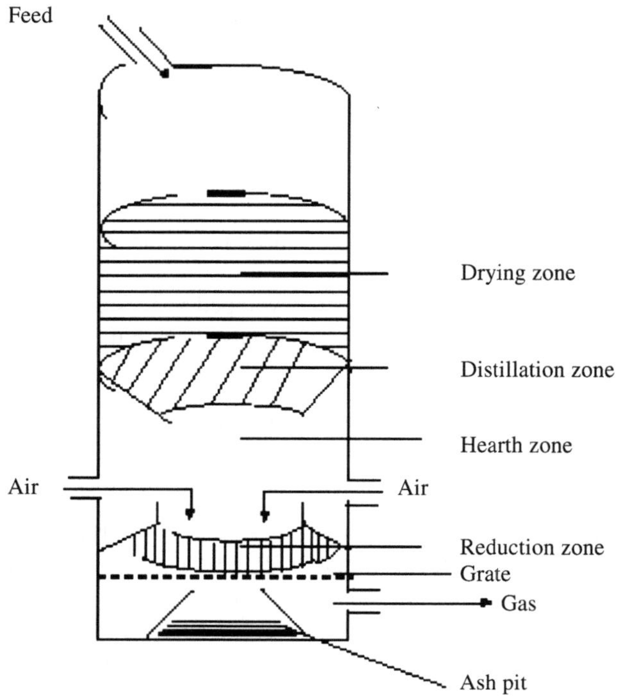

Fig. 6.19 Schematic diagram of a downdraft fixed-bed gasifier

draft) to 20% (updraft) or greater in the product gases. The oxidation reactions of the volatiles are very rapid and the oxygen is consumed before it can diffuse to the surface of the char. In the reduction zone (often referred to as the gasification zone) the char is converted into product gas by reaction with the hot gases from the upper zones. Depending on the end use, it is necessary to cool and clean the gas in order to remove as much water vapor, dust and pyrolytic products as possible from the gas, especially if it is to be used in an internal combustion engine.

Fixed-bed gasifiers are the most suitable for biomass gasification. Fixed-bed gasifiers are usually fed from the top of the reactor and can be designed in either updraft or downdraft configurations. The product gases from these two gasifier configurations vary significantly. At larger scales, fixed-bed gasifiers can encounter problems with bridging of the biomass feedstock. This leads to uneven gas flow. Achieving uniform temperatures throughout the gasifier on large scales can also be difficult due to the absence of mixing in the reaction zone. Most fixed-bed gasifiers are air-blown and produce low-energy gases (Stevens 2001).

With fixed-bed updraft gasifiers, the air or oxygen passes upward through a hot reactive zone near the bottom of the gasifier in a direction counter-current to the flow of solid material. Exothermic reactions between air/oxygen and the charcoal in the bed drive the gasification process. Heat in the raw gas is transferred to the biomass feedstock as the hot gases pass upward, and biomass descending through the gasifier sequentially undergoes drying, pyrolysis, and finally gasification. Fixed-bed updraft gasifiers can be scaled-up; however, they produce a product gas with very high tar concentrations. This tar should be removed for the major part from the gas, creating a gas-cleaning problem.

Fixed-bed downdraft gasifiers were widely used in World War II for operating vehicles and trucks. During operation, air is drawn downward through a fuel bed; the gas in this case contains relatively less tar compared with the other gasifier types. Fixed-bed downdraft gasifiers are limited in scale and require a well-defined fuel, making them not fuel-flexible.

Fluidized-bed (FB) gasifiers are a more recent development that take advantage of the excellent mixing characteristics and high reaction rates of this method of gas-solid contacting. Examples of the FB gasifier systems are the bubbling fluidized-bed (BFB) gasifiers, the entrained-bed (EB) gasifiers, and the circulating fluidized-bed (CFB) gasifiers.

The FB gasifiers are typically operated at 1050–1250 K (limited by the melting properties of the bed material) and are therefore not generally suitable for coal gasification, as due to the lower reactivity of coal compared to biomass, and higher required temperature (>1550 K). Heat to drive the gasification reaction can be provided in a variety of ways in FB gasifiers. Direct heating occurs when air or oxygen in fluidizing gas partially oxidizes the biomass and heat is released by the exothermic reactions that occur. Indirect heating methods such as internal heat exchangers, using preheated bed material, or other means can also be used to drive the gasification reactions (Stevens 2001). The BFB gasifier tends to produce a gas with tar content between that of the updraft and downdraft gasifiers. Some pyrolysis products are swept out of the fluid bed by gasification products, but are then further converted by thermal cracking in the freeboard region (Warnecke 2000). The CFB gasifiers employ a system where the bed material circulates between the gasifier and a secondary vessel. The CFB gasifiers are suitable for fuel capacity higher than 10 MWth (Susta et al. 2003). The FB gasifier, and the CFB gasifier systems are given in Fig. 6.11.

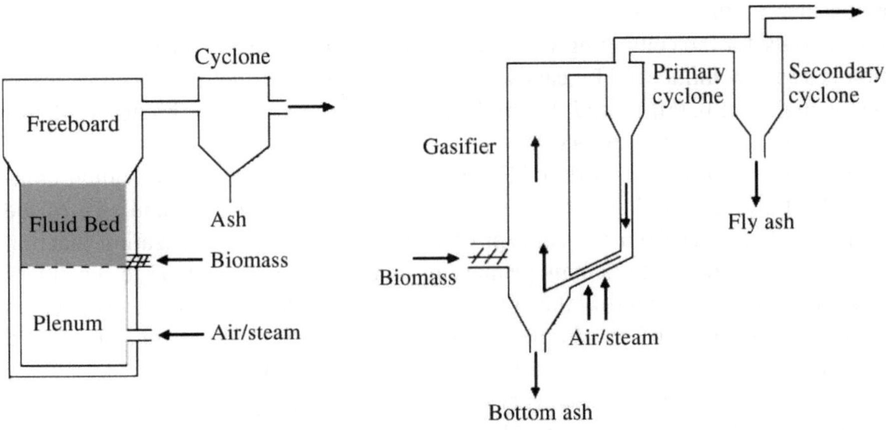

Fluidized Bed Gasifier Circulating Fluidized Bed Gasifier

Fig. 6.20 Fluidized-bed and circulating fluidized-bed gasifier systems

6.5.2 *Integrated Gasification-Fischer–Tropsch (BIG-FT) Process*

The Fischer–Tropsch synthesis (FTS) was established in 1923 by German scientists Franz Fischer and Hans Tropsch. The main aim of FTS is synthesis of long-chain hydrocarbons from a CO and H_2 gas mixture. The FTS is described by (Schulz 1999):

$$nCO + (n + m/2)\,H_2 \rightarrow C_nH_m + nH_2O \tag{6.11}$$

In the FTS one mole of CO reacts with two moles of H_2 in the presence of cobalt (Co) based catalyst to afford a hydrocarbon chain extension ($-CH_2-$). The reaction of synthesis is exothermic ($\Delta H = -165\,kJ/mol$):

$$CO + 2H_2 \rightarrow -CH_2- + H_2O \qquad \Delta H = -165\,kJ/mol \tag{6.12}$$

The $-CH_2-$ is a building block for longer hydrocarbons. One main characteristic regarding the performance of the FTS is the liquid selectivity of the process (Stelmachowski and Nowicki 2003).

The FTS is a well-established process for the production of synfuels. The process is used commercially in South Africa by Sasol and Mossgas and in Malaysia by Shell. The FTS can be operated at low temperatures (LTFT) to produce a syncrude with a large fraction of heavy, waxy hydrocarbons or it can be operated at higher temperatures (HTFT) to produce a light syncrude and olefins. With HTFT the primary products can be refined to environmentally friendly gasoline and diesel, solvents and olefins. With LTFT, the heavy hydrocarbons can be refined to waxes or if hydrocracked and/or isomerized, to produce excellent diesel, base stock for lube oils and a naphtha that is ideal feedstock for cracking to light olefins. Biomass integrated gasification-Fischer–Tropsch (BIG-FT) process is favorable as a production route.

The FTS has been widely investigated for more than 70 years, and Fe and Co are typical catalysts. Cobalt-based catalysts are preferred because their productivity is better than Fe thanks to their high activity, selectivity for linear hydrocarbons, and low activity for the competing water–gas shift reaction. The FTS product composition is strongly influenced by catalyst composition: the product from a cobalt catalyst is higher in paraffins and the product from an iron catalyst is higher in olefins and oxygenates (Andersen 1984).

The variety of composition of FT products with hundreds of individual compounds shows a remarkable degree of order with regard to class and size of the molecules. Starting from the concept of FTS as an ideal polymerization reaction it is easily realized that the main primary products, olefins, can undergo secondary reactions and thereby modify the product distribution. This generally leads to chain length dependencies of certain olefin reaction possibilities, which are again suited to serve as a characteristic feature for the kind of olefin conversion.

While gasification processes vary considerably, typically gasifiers operate from 975 K and higher and from atmospheric pressure to 5 atm or higher. The process is generally optimized to produce fuel or feedstock gases. Gasification processes also produce a solid residue as a char, ash, or slag. The product fuel gases, includ-

Table 6.19 Typical composition of the syngas

Gaseous product	Percentage by volume
Hydrogen	29–40
Carbon monoxide	21–32
Methane	10–15
Carbon dioxide	15–20
Ethylene	0.4–1.2
Water vapor	4–8
Nitrogen	0.6–1.2

ing hydrogen, can be used in internal and external combustion engines, fuel cells, and other prime movers for heat and mechanical or electrical power. Gasification products can be used to produce methanol, FT liquids, and other fuel liquids and chemicals. The typical composition of the syngas is given in Table 6.19.

Catalytic steam reforming of hydrocarbons has been extensively studied, especially in the context of methane reforming to make syngas (H_2:CO = 2:1) for methanol and FT liquid synthesis. Methanol is produced in large quantities by the conversion of natural gas or petroleum into a synthesis gas. Biomass has also been used as a feedstock for producing synthesis gas used in production of both methanol and FT liquids.

In all types of gasification, biomass is thermochemically converted to a low or medium-energy content gas. The higher heating value of syngas produced from biomass in the gasifier is typically 10–13 MJ/Nm^3. Air-blown biomass gasification results in approximately 5 MJ/Nm^3 and oxygen-blown 15 MJ/Nm^3 of gas and is considered a low to medium energy content gas compared to natural gas (35 MJ/Nm^3).

The processing of synfuels from biomass will lower the energy cost, improve the waste management, and reduce harmful emissions. This triple assault on plant operating challenges is a proprietary technology that gasifies biomass by reacting it with steam at high temperatures to form a clean burning syngas. The molecules in the biomass (primarily carbon, hydrogen and oxygen) and the molecules in the steam (hydrogen and oxygen) reorganize to form this syngas.

Reforming the light hydrocarbons and tars formed during biomass gasification also produces hydrogen. In essence, the system embodies a fast, continuous process for pyrolizing or thermally decomposing biomass and steam reforming the resulting constituents. The entire process occurs in reducing environment biomass gasifiers. Steam reforming and so-called dry or CO_2 reforming occur according to the following reactions and are usually promoted by the use of catalysts.

$$C_nH_m + nH_2O \leftrightarrows n\,CO + (n + m/2)\,H_2 \tag{6.13}$$

$$C_nH_m + nCO_2 \leftrightarrows (2n)\,CO + (m/2)\,H_2 \tag{6.14}$$

The high-temperature FT technology applied by Sasol in the synthol process in South Africa at the Secunda petrochemical site is the largest commercial scale application of FT technology. The most recent version of this technology is the

Sasol Advanced Synthol (SAS) process. Although the FTS was initially envisioned as a means to make transportation fuels, there has been a growing realization that the profitability of commercial operations can be improved by the production of chemicals or chemical feedstock. FTS yields a complex mixture of saturated or unsaturated hydrocarbons (C_1-C_{40+}), C_1-C_{16+} oxygenates as well as water and CO_2. Laboratory studies are often carried out in differential conditions (conversions of approx. 3%) and therefore very accurate product analysis is necessary. Hydrocarbons obtained from a SAS type reactor are: for C_5-C_{10} and $C_{11}-C_{14}$ hydrocarbons, paraffins 13 and 15%, olefins 70 and 60%, aromatics 5 and 15%, and oxygenates 12 and 10%, respectively. From the component breakdown of the main liquid cuts it is clear that there is considerable scope for producing chemical products in addition to hydrocarbon fuels (Anderson 1984; Schulz 1999; Dry 2004).

The FTS in supercritical phase can be developed and its reaction behavior and mass transfer phenomenon were analyzed by both experimental and simulation methods. The same reaction apparatus and catalysts can be used for both gas phase reaction and liquid phase FT reactions to correctly compare characteristic features of diffusion dynamics and the reaction itself, in various reaction phases. Efficient transportation of reactants and products to the inside of catalysts bed and pellet, quick heat transfer, and in situ product extraction from catalysts by supercritical fluid can be accomplished.

There has been an increasing interest in the effect of water on cobalt FT catalysts in recent years. Water is produced in large amounts over cobalt catalysts since one water molecule is produced for each C-atom added to a growing hydrocarbon chain and due to the low water–gas shift activity of cobalt. The presence of water during FTS may affect the synthesis rate reversibly as reported for titania-supported catalysts, the deactivation rate as reported for alumina-supported catalysts and water also has a significant effect on the selectivity for cobalt catalysts on different supports. The effect on the rate and the deactivation appears to depend on the catalyst system studied while the main trends in the effect on selectivity appear to be more consistent for different supported cobalt systems. There are, however, also some differences in the selectivity effects observed. The present study deals mainly with the effect of water on the selectivity of alumina-supported cobalt catalysts but some data on the activity change will also be reported. The results will be compared with results for other supported cobalt systems reported in the literature (Dry 2004).

The activity and selectivity of supported Co FTS catalysts depends on both the number of Co surface atoms and on their density within support particles, as well as on transport limitations that restrict access to these sites. Catalyst preparation variables that are available to modify these properties include cobalt precursor type and loading level, support composition and structure, pretreatment procedures, and the presence of promoters or additives. Secondary reactions can strongly influence product selectivity. For example, the presence of acid sites can lead to the useful formation of branched paraffins directly during the FTS step. However, product water not only oxidizes Co sites making them inactive for additional turnovers, but it can inhibit secondary isomerization reactions on any acid sites intentionally placed in FTS reactors (Riedel et al. 1999).

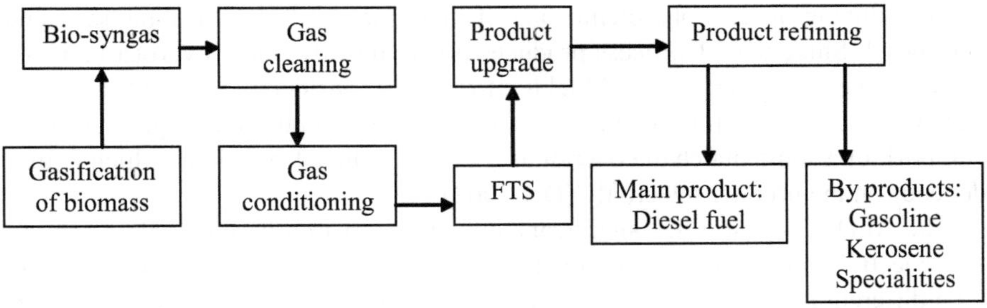

Fig. 6.21 Production of diesel fuel from biosyngas by Fischer–Tropsch synthesis (FTS)

Iron catalysts used commercially by Sasol in the Fischer–Tropsch synthesis for the past five decades (Dry 2002) have several advantages: (1) lower cost relative to cobalt and ruthenium catalysts, (2) high water–gas shift activity allowing utilization of syngas feeds of relatively low hydrogen content such as those produced by gasification of coal and biomass, (3) relatively high activity for production of liquid and waxy hydrocarbons readily refined to gasoline and diesel fuels, and (4) high selectivity for olefinic C_2–C_6 hydrocarbons used as chemical feedstocks. The typical catalyst used in fixed-bed reactors is an unsupported Fe/Cu/K catalyst prepared by precipitation (Wang et al. 2003). While having the previously mentioned advantages, this catalyst (1) deactivates irreversibly over a period of months to a few years by sintering, oxidation, formation of inactive surface carbons, and transformation of active carbide phases to inactive carbide phases, and (2) undergoes attrition at unacceptably high rates in the otherwise highly efficient, economical slurry bubble-column reactor.

It is well known that addition of alkali to iron causes an increase of both the 1-alkene selectivity and the average carbon number of produced hydrocarbons. While the promoter effects on iron has been thoroughly studied only few and on a first glance contradictive results are available for cobalt catalysts. In order to complete experimental data the carbon number distributions are analyzed for products obtained in a fixed-bed reactor under steady state condition. Precipitated iron and cobalt catalysts with and without K_2CO_3 were used (Riedel et al. 1999; Dry 2002).

Figure 6.21 shows the production of diesel fuel from biosyngas by FTS. The design of a biomass gasifier integrated with a FTS reactor must be aimed at achieving a high yield of liquid hydrocarbons. For the gasifier, it is important to avoid methane formation as much as possible, and convert all carbon in the biomass to mainly carbon monoxide and carbon dioxide (Prins et al. 2004).

The FTS-based gas to liquids technology (GTL) includes the three processing steps namely syngas generation, syngas conversion and hydroprocessing. In order to make the GTL technology more cost-effective, the focus must be on reducing both the capital and the operating costs of such a plant (Vosloo 2001). For some time now the price has been up to \$60 per barrel. It has been estimated that the FT process should be viable at crude oil prices of about \$20 per barrel (Jager 1998). The current commercial applications of the FT process are geared at the production of the valuable linear alpha olefins and of fuels such as LPG, gasoline, kerosene and diesel. Since the FT process produces predominantly linear hydrocarbons the production of high-quality diesel fuel is currently of considerable interest. The most expensive section of an FT complex is the production of purified syngas and so its composition should match the overall usage ratio of the FT reactions, which in turn depends on the product selectivity (Dry 2002).

The Al_2O_3/SiO_2 ratio has significant influences on iron-based catalyst activity and selectivity in the process of FTS. Product selectivities also change significantly with different Al_2O_3/SiO_2 ratios. The selectivity of low molecular weight hydrocarbons increases and the olefin to paraffin ratio in the products shows a monotonic decrease with increasing Al_2O_3/SiO_2 ratio. Recently, Jun et al. (2004) studied FTS over Al_2O_3 and SiO_2 supported iron-based catalysts from biomass-derived syngas. They found that Al_2O_3 as a structural promoter facilitated the better dispersion of copper and potassium and gave much higher FTS activity.

Recently, there has been some interest in the use of FTS for biomass conversion to synthetic hydrocarbons. Biomass can be converted to biosyngas by non-catalytic, catalytic and steam gasification processes. The biosyngas consists mainly of H_2, CO, CO_2 and CH_4. The FTS has been carried out using $CO/CO_2/H_2/Ar$ (11/32/52/5 vol.%) mixture as a model for biosyngas on co-precipitated Fe/Cu/K, Fe/Cu/Si/K and Fe/Cu/Al/K catalysts in a fixed-bed reactor. Some performances of the catalysts that depended on the syngas composition are also presented (Jun et al. 2004). The kinetic model predicting the product distribution is taken from Wang et al. (2003), for an industrial Fe/Cu/K catalyst.

To produce biosyngas from a biomass fuel the following procedures are necessary: (a) gasification of the fuel, (b) cleaning of the product gas, (c) usage of the synthesis gas to produce chemicals, and (d) usage of the synthesis gas as energy carrier in fuel cells. Figure 6.22 shows the green diesel and other products from biomass via Fischer–Tropsch synthesis.

6.5.3 Upgrading of Gasification Products

Main biorenewable gaseous fuels are biogas, landfill gas, gaseous fuels from pyrolysis and gasification of biomass, gaseous fuels from Fischer–Tropsch synthesis and biohydrogen (Balat 2008; Balat 2009). Figure 6.23 shows biomass upgrading by gasification and anaerobic digestion. There are a number of processes for converting of biomass into gaseous fuels such as methane or hydrogen. One pathway uses plant and animal wastes in a fermentation process leading to biogas from which the desired fuels can be isolated. This technology is established and in widespread use for waste treatment. Anaerobic digestion of biowastes occurs in the absence of air, the resulting gas known as biogas is a mixture consisting mainly of methane and carbon dioxide. Biogas is a valuable fuel, which is produced in digesters filled with the feedstock like dung or sewage. The digestion is allowed to continue for a period from ten days to a few weeks. A second pathway uses algae and bacteria that have been genetically modified to produce hydrogen directly instead of the conventional biological energy carriers. Finally, high-temperature gasification supplies a crude

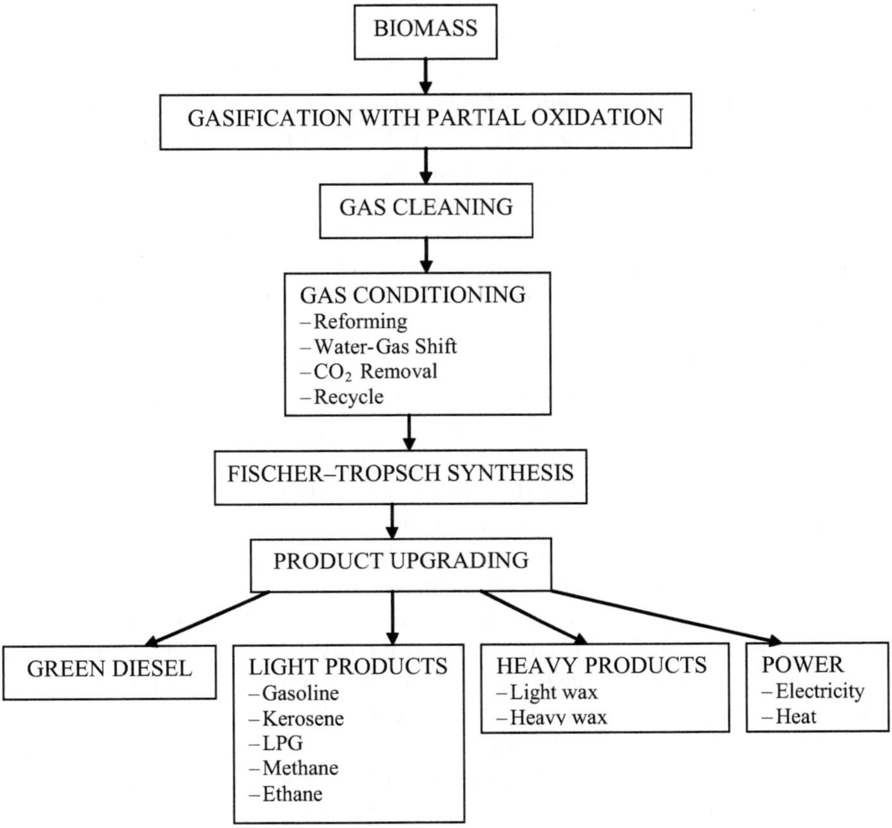

Fig. 6.22 Green diesel and other products from biomass via Fischer–Tropsch synthesis

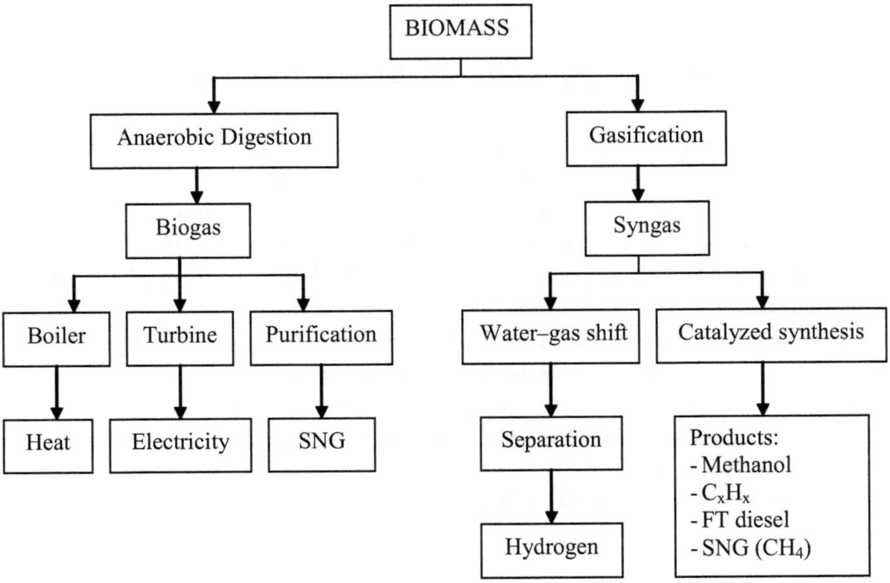

Fig. 6.23 Biomass upgrading by gasification and anaerobic digestion

gas, which may be transformed into hydrogen by a second reaction step. This pathway may offer the highest overall efficiency.

Hydrogen can be produced from biomass via two thermochemical processes: (1) gasification followed by reforming of the syngas, and (2) fast pyrolysis followed by reforming of the carbohydrate fraction of the bio-oil. In each process, a water–gas shift is used to convert the reformed gas into hydrogen, and pressure swing adsorption is used to purify the product. Gasification technologies provide the opportunity to convert biorenewable feedstocks into clean fuel gases or synthesis gases. The synthesis gas includes mainly hydrogen and carbon monoxide (H_2+CO), which is also called syngas. Biosyngas is a gas rich in CO and H_2 obtained by gasification of biomass.

Hydrogen can be produced from biomass by pyrolysis, gasification, steam gasification, steam reforming of bio-oils, and enzymatic decomposition of sugars. Hydrogen is produced from pyroligneous oils produced from the pyrolysis of lignocellulosic biomass. The yield of hydrogen that can be produced from biomass is relatively low, 16–18% based on dry biomass weight (Demirbas 2001).

The strategy is based on producing hydrogen from biomass pyrolysis using a co-product strategy to reduce the cost of hydrogen and concluded that only this strategy could compete with the cost of the commercial hydrocarbon-based technologies (Wang et al. 1998). This strategy will demonstrate how hydrogen and biofuel are economically feasible and can foster the development of rural areas when practiced on a larger scale. The process of biomass to activated carbon is an alternative route to hydrogen with a valuable co-product that is practiced commercially. The yield of hydrogen that can be produced from biomass is relatively low, 12–14% based on the biomass weight (Demirbas 2005). In a proposed second process, fast pyrolysis of biomass is used to generate bio-oil and catalytic steam reforming of the bio-oil produces hydrogen and carbon dioxide.

6.5.4 Refining of Upgrading Gasification Products

Thermochemical and biochemical conversion products from biomass are upgraded before ultimate refining processes. Biorefinery includes fractionation for separation of primary refinery products. The main goals of biorefineries are to integrate production of higher value chemicals and commodities, as well as fuels and energy, and to optimize use of resources, maximize profitability, maximize benefits and minimize wastes.

The benefits of an integrated biorefinery are numerous because of the diversification in feedstocks and products. There are currently several different levels of integration in biorefineries, which adds to their sustainability, both economically and environmentally. Economic and production advantages increase with the level of integration in the biorefinery.

Biorefineries can be classified into the following categories accordingly their functions:

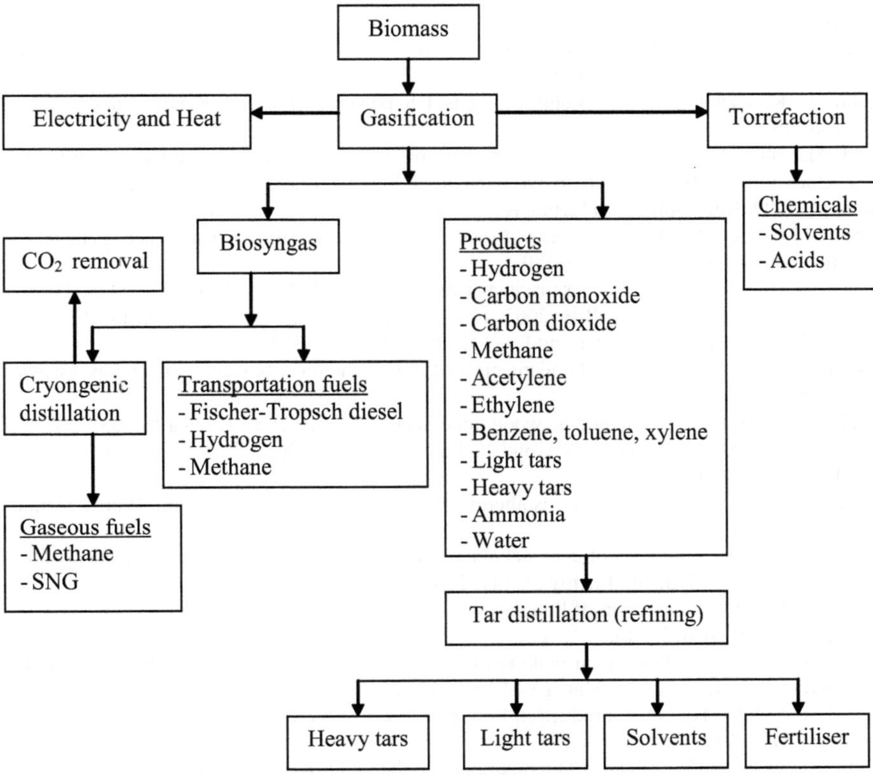

Fig. 6.24 Gasification-based thermochemical biorefinery

- Fast pyrolysis-based biorefineries
- Gasification-based biorefineries
- Sugar-based biorefineries
- Green biorefinery
- Energy crops biorefinery
- Oilseed biorefinery
- Forest-based and lignocellulosic biorefinery

Figure 6.24 shows the gasification-based thermochemical biorefinery.

References

An, J., Bagnell, L., Cablewski, T., Strauss, C.R., Trainor, R.W. 1997. Applications of high temperature aqueous media for synthetic organic reactions. J Org Chem 62:2505–2511.

Anderson, R.B. 1984. The Fischer–Tropsch synthesis. Academic, New York.

Appell, H.R., Fu, Y.C., Friedman, S., Yavorsky, P.M., Wender, I. 1971. Converting organic wastes to oil. US Bureau of Mines Report of Investigation. No. 7560.

Appell, H.R. 1967. In: Fuels from waste, Anderson, L., Tillman, D.A. (Eds.) Academic, New York.
Babu, B.V., Chaurasia, A.S. 2003. Modeling for pyrolysis of solid particle: kinetics and heat transfer effects. Energy Convers Manage 44:2251–2275.
Badin, J., Kirschner, J. 1998. Biomass greens US power production. Renew Energy World 1:40-45.
Balat, M. 2008. Hydrogen-rich gas production from biomass via pyrolysis and gasification processes and effects of catalyst on hydrogen yield. Energy Sources Part A 30:552–564.
Balat, M. 2009. New biofuel production technologies. Energy Edu Sci Technol 22:147–161.
Barooah, J.N., Long, V.D. 1976. Rates of thermal decomposition of some carbonaceous materials in a fluidized bed. Fuel 55:116–120.
Beaumont, O. 1985. Flash pyrolysis products from beech wood. Wood Fiber Sci. 17:228–239.
Byrd, A.J., Pant, K.K., Gupta, R.B. 2007. Hydrogen production from glucose using Ru/Al_2O_3 catalyst in supercritical water. Ind Eng Chem Res 46:3574–3579.
Chornet, E., Overend, R.P.1985. Fundamentals of thermochemical biomass conversion. Elsevier Applied, New York.
Demirbas, A. 1985. A new method on wood liquefaction. Chimica Acta Turcica 13:363–368.
Demirbas, A. 1992. Conversion of wood to liquid products using alkaline glycerol. Fuel Sci Technol Int 10:173–184.
Demirbas, A. 1998. Kinetics for non-isothermal flash pyrolysis of hazelnut shell. Biores Technol 66:247–252.
Demirbas, A. 1999. Properties of charcoal derived from hazelnut shell and the production of briquets using pyrolytic oil. Energy 24:141–150.
Demirbas, A. 2000. Mechanism of liquefaction and pyrolysis reactions of biomass. Energy Convers Manage 41:633–646.
Demirbas, A. 2001. Carbonization ranking of selected biomass for charcoal, liquid and gaseous products. Energy Convers Manage 42:1229–1238.
Demirbas, A. 2005. Hydrogen production from biomass via supercritical water extraction. Energy Sources 27:1409–1417.
Demirbas, A. 2007. The influence of temperature on the yields of compounds existing in bio-oils obtaining from biomass samples via pyrolysis. Fuel Proc Technol 88:591–597.
Demirbas, A. 2008a. Conversion of corn stover to chemicals and fuels. Energy Sources Part A 30:788–796.
Demirbas, A. 2008b. New liquid biofuels from vegetable oils via catalytic pyrolysis. Energy Edu Sci Technol 21:1–59.
Desrosiers, R.E., Lin, R.J. 1984. A moving-boundary model of biomass pyrolysis. Solar Energy 33:187–196.
Dry, M.E. 2002. The Fischer–Tropsch process: 1950–2000. Catal Today 71:227–241.
Dry, M.E. 2004. Present and future applications of the Fischer–Tropsch process. Appl Catal A General 276:1–3.
Feng, W., van der Kooi, H.J., Arons, J.D.S. 2004. Biomass conversions in subcritical and supercritical water: driving force, phase equilibria, and thermodynamic analysis. Chem Eng Proc 43:1459–1467.
Gadhe, J.B., Gupta, R.B. 2007. Hydrogen production by methanol reforming in supercritical water: catalysis by in situ-generated copper nanoparticles. Int J Hydrogen Energy 32:2374–2381.
Goudriaan, F., Peferoen, D. 1990. Liquid fuels from biomass via a hydrothermal process. Chem Eng Sci 45:2729–2734.
Hao, X.H., Guo, L.J., Zhang, X.M., Guan, Y. 2005. Hydrogen production from catalytic gasification of cellulose in supercritical water. Chem Eng J 110:57–65.
Hofmann, L., Antal, M.J. 1984. Numerical simulations of the performance of solar fired flash pyrolysis reactors. Solar Energy 33:427–440.
Hsu, C.-C., Hixon, A.N. 1981. C1 to nC4 oxygenated compounds by promoted pyrolysis of cellulose. Ind Eng Chem Prod Res Develop 20:109–114.
Inoue, S., Sawayma, S., Dote, Y., Ogi, T. 1997. Behavior of nitrogen during liquefaction of dewatered sewage sludge. Biomass Bioenergy 12:473–475.

Itoh, S., Suzuki, A., Nakamura, T., Yokoyama, S. 1994. Production of heavy oil from sewage sludge by direct thermochemical liquefaction. In: Proceedings of the IDA and WRPC World Congress on Desalination and Water Treatment, 1994.

Jager, B. 1998. In: Proceedings of the 5th Natural Gas Conversion Symposium, Taormina, Italy, 1998.

Jun, K.W., Roh, H.S., Kim, K.S., Ryu, J.S., Lee, K.W. 2004. Catalytic investigation for Fischer–Tropsch synthesis from bio-mass derived syngas. Appl Catal A 259:221–226.

Kaparaju, P., Serrano, M., Thomsen, A.B., Kongjan, P., Angelidaki, I. 2009. Bioethanol, biohydrogen and biogas production from wheat straw in a biorefinery concept. Biores Technol 100:2562–2568.

Koullas, D.P., Nikolaou, N., Koukkios, E. g. 1998. Modelling non-isothermal kinetics of biomass prepyrolysis at low pressure. Biores Technol 63:261–266.

Kranich, W.L. 1984. Conversion of sewage sludge to oil by hydroliquefaction. EPA-600/2 84-010. Report for the U.S. Environmental Protection Agency. EPA, Cincinnati, OH.

Kruse, A., Meier, D., Rimbrecht, P., Schacht, M. 2000. Gasification of pyrocatechol in supercritical water in the presence of potassium hydroxide. Ind Eng Chem Res 39:4842–2848.

Kucuk, M.M. 2005. Delignification of biomass using alkaline glycerol. Energy Sources 27:1245–1255.

Kucuk, M.M., Demirbas, A. 1993. Delignification of Ailanthus altissima and Spruce orientalis with glycerol or alkaline glycerol at atmospheric pressure. Cellulose Chem Technol 27:679–686.

Kuhlmann, B., Arnett, E.M., Siskin, M. 1994. Classical organic reactions in pure superheated water. J Org Chem 59:3098–3101.

Midgett, J.S. 2008. Assessing a hydrothermal liquefaction process using biomass feedstocks. Master's Thesis, Louisiana State University, Baton Rouge, LA.

Minowa, T., Ogi, T., Dote, Y., Yokoyama, S. 1994. Effect of lignin content on direct liquefaction of bark. Int Chem Eng 34:428–430.

Minowa, T., Zhen, F., Ogi, T. 1997. Cellulose decomposition in hot-compressed water with alkali or nickel catalyst. J Supercritical Fluids 13:253–259.

Mohan, D., Pittman, C.U., Steele, P.H. 2006. Pyrolysis of wood/biomass for bio-oil: a critical review. Energy Fuels 20:848–889.

Molten, P.M., Demmitt, T.F., Donovan, J.M., Miller, R.K. 1983. Mechanism of conversion of cellulose wastes to liquid in alkaline solution. In: Energy from biomass and wastes. Klass, D.L. (Ed.) Institute of Gas Technology, Chicago.

Ogi, T., Yokoyama, S., Koguchi, K. 1985. Direct liquefaction of wood by alkali and alkaline earth salt in an aqueous phase. Chem Lett 8:1199–1202.

Ogi, T., Yokoyama S. 1993. Liquid fuel production from woody biomass by direct liquefaction. Sekiyu Gakkaishi 36:73–84.

Overend, R.P. 1998. Biomass gasification: a growing business. Renew Energy World 1:59–63.

Peterson, A.A., Vogel, F., Lachance, R.P., Froling, M., Antal, M.J., Tester, J.W. 2008. Thermochemical biofuel production in hydrothermal media: a review of sub- and supercritical water technologies. Energy Environ Sci 1:32–65.

Prins, M.J., Ptasinski, K.J., Janssen, F.J.J.G. 2004. Exergetic optimisation of a production process of Fischer–Tropsch fuels from biomass. Fuel Proc Technol 86:375–389.

Riedel, T., Claeys, M., Schulz, H., Schaub, G., Nam, S.S., Jun, K.W., Choi, M.J., Kishan, G., Lee, K.W. 1999. Comparative study of FTS with H_2/CO and H_2/CO_2 syngas using Fe and Co catalysts. Appl Cat A 186:201–213.

Russell, J.A., Miller, R.K., Molten, P.M. 1983. In: Biomass, Coombs, J., Hall, D.O. (Eds.) Biomass 3:43–57.

Sang, O.Y., Twaiq, F., Zakaria, R., Mohamed, A., Bhatia, S. 2003. Biofuel production from catalytic cracking of palm oil. Energy Sources 25:859–869.

Sasaki, M., Kabyemela, B.M., Malaluan, R.M., Hirose, S., Takeda, N., Adschiri, T., Arai. K. 1998. Cellulose hydrolysis in subcritical and supercritical water. J Supercritical Fluids 13:261–268.

Scott, D.S., Paterson, L., Piskorz, J., Radlein, D. 2000. Pretreatment of poplar wood for fast pyrolysis: rate of cation removal. J Anal Appl Pyrolysis 57:169–176.

Schulz, H. 1999. Short history and present trends of FT synthesis. Appl Catal A General 186:1–16.

Stamm, A.J. 1956. Thermal degradation of wood and cellulose. Ind Eng Chem 48:413–417.

Stelmachowski, M., Nowicki, L. 2003. Fuel from the synthesis gas: the role of process engineering. Appl Energy 74:85–93.

Stevens, D.J. 2001. Hot gas conditioning: recent progress with larger-scale biomass gasification systems. National Renewable Energy Laboratory, NREL/SR-510-29952. National Renewable Energy Laboratory, Golden, CO.

Susta, M.R., Luby, P., Mat, S.B. 2003. Biomass energy utilization and environment protection commercial reality and outlook, Power Gen Asia. http://www.powergeneration.siemens.com/download/pool/industrialheatpower_02.pdf. Accessed 2009.

Suzuki, A., Yokoyama, S., Murakami, M., Ogi, T., Koguchi, K. 1986. New treatment of sewage sludge by direct thermochemical liquefaction. Chem Lett 9:1425–1428.

Timell, T.E. 1967. Recent progress in the chemistry of wood. Hemicelluloses. Wood Sci Technol 1:45–70.

Tran, D.Q., Charanjit, R. 1978. A kinetic model for pyrolysis of Douglas fir bark. Fuel 57:293–298.

Vosloo, A.C. 2001. Fischer–Tropsch: a futuristic view. Fuel Proc Technol 71:149–155.

Wang, D., Czernik, S., Chornet, E. 1998. Production of hydrogen from biomass by catalytic steam reforming of fast pyrolysis oils. Energy Fuels 12:19–24.

Wang, Y.N., Ma, W.P., Lu, Y.J., Yang, J., Xu, Y.Y., Xiang, H.W., Li, Y.W., Zhao, Y.L., Zhang, B.J. 2003. Kinetics modeling of FT synthesis over an industrial Fe-Cu-K catalyst. Fuel 82:195–213.

Warnecke, R. 2000. Gasification of biomass: comparison of fixed bed and fluidized bed gasifier. Biomass Bioenergy 18:489–497.

Zhang, Y. 2008. Reviving the carbohydrate economy via multi-product lignocellulose biorefineries. J Ind Microbiol Biotechnol 35:367–375.

Zhang, S.P., Yan, Y.J., Ren, J.W., Li, T.C. 2003. Study of hydrodeoxygenation of bio-oil from the fast pyrolysis of biomass. Energy Sources 25:57–65.

Chapter 7
Biochemical Processes

7.1 Introduction

Biochemical conversion proceeds at lower temperatures and lower reaction rates and can offer high selectivity for products. Higher moisture feedstocks are generally good candidates for biochemical processes. Ethanol production is a biochemical conversion technology used to produce energy from biomass. Many highly efficient biochemical conversion technologies have developed in nature to break down the molecules of which biomass is composed. For ethanol production, biochemical conversion researchers have focused on a process model of dilute acid hydrolysis of hemicelluloses followed by enzymatic hydrolysis of cellulose. Biodiesel production is a biochemical conversion technology used to produce energy from oilseed crops (Shieh et al. 2003; Du et al. 2004; Demirbas 2005; Noureddini et al. 2005; Demirbas 2008a; Demirbas 2009).

Chemical energy and organic carbon are obtained by organisms either directly or indirectly via the photosynthetic conversion of solar energy. These organisms have evolved metabolic machineries for the photochemical reduction of carbon dioxide to organic matter. The majority of the bioengineering strategies for biochemically derived fuels involve options for the disposition of organic matter produced via photosynthate. The bulk of the presently exploited photosynthate is directed toward the production of wood, food, and feed. During processing and consumption, waste organic materials are generated which can be used for energy production via combustion, pyrolysis or biochemical conversions to ethanol, hydrogen, methane, and iso-propanol.

7.2 Biochemical Conversion Facilities

Ethanol is the most widely used liquid biofuel. It is an alcohol and is fermented from sugars, starches or from cellulosic biomass. Cellulosic materials can be used to produce bioethanol. Bioethanol represents an important, renewable liquid fuel for motor vehicles. Production of bioethanol from biomass is one way to reduce both the consumption of crude oil and environmental pollution. Conversion technologies for producing ethanol from cellulosic biomass resources such as forest materials, agricultural residues and urban wastes are under development and have not yet been demonstrated commercially. In order to produce bioethanol from cellulosic biomass, a pretreatment process is used to reduce the sample size, break down the hemicelluloses to sugars, and open up the structure of the cellulose component. The cellulose portion is hydrolyzed by acids or enzymes into glucose sugar that is fermented to bioethanol. The sugars from the hemicelluloses are also fermented to bioethanol. Bioethanol has long been used as a motor fuel. It began with the use of ethanol in the internal combustion engine (Demirbas 2005).

Fermentation feedstocks require pretreatment by chemical, physical, or biological means to open up the structure of biomass and reduce the complex carbohydrates to simple sugars. This set of pretreatments is often referred to as hydrolysis. The resulting sugars can then be fermented by the yeast and bacteria employed in the process. Feedstocks high in starch and sugar are most easily hydrolyzed. Cellulosic feedstocks, including the major fraction of organics in MSW, are more difficult to hydrolyze, requiring more extensive pretreatment.

7.3 Production of Fuels

Global production of bioethanol reached 13.5 billion gallons in 2006 (Demirbas 2008b). Bioethanol currently accounts for more than 94% of global biofuel production, with the majority coming from sugarcane. About 60% of global bioethanol production comes from sugarcane and 40% from other crops. Brazil and the US are the world leaders, which together account for about 70% of the world bioethanol production, exploiting both sugarcane and corn. Ethanol has been used in Germany and France as early as 1894 by the then incipient industry of internal combustion engines. By that time, the production of ethanol was 70 times greater than the production and consumption of petrol. Currently, ethanol is produced from sugar beets and from molasses. A typical yield is 72.5 L of ethanol per ton of sugar cane. Production of ethanol from biomass is one way to reduce both the consumption of crude oil and environmental pollution. The use of gasohol (an ethanol and gasoline mixture) as an alternative motor fuel has been steadily increasing in the world for a number of reasons. For example, domestic production and use of ethanol for fuel can decrease dependence on foreign oil, reduce trade deficits, create jobs in rural areas, reduce air pollution, and reduce global climate change and carbon dioxide buildup (Bala 2005).

Ethanol from lignocellulosic biomass has the potential to contribute substantially to bioethanol for transportation. In the process evaluated, prehydrolysis with dilute sulfuric acid is employed to hydrolyze hemicellulose and make the cellulose more accessible to hydrolysis by enzymes. Residual biomass from hydrolysis and extraction of carbohydrates can be burned in a power plant to generate electricity and process steam. Figure 7.1 shows a flow diagram of ethanol production from sugar crops and lignocellulosic feedstocks.

Carbohydrates (hemicelluloses and cellulose) in plant materials can be converted to sugars by hydrolysis process. Fermentation is an anaerobic biological process in which sugars are converted to alcohol by the action of microorganisms, usually yeast. The resulting alcohol is bioethanol. The value of any particular type of biomass as feedstock for fermentation depends on the ease with which it can be converted to sugars. Bioethanol is a petrol additive/substitute. Bioethanol and the biorefinery concept are closely linked. It is possible that wood, straw and even household wastes may be economically converted to bioethanol. In 2004, 3.4 billion gallons of fuel ethanol were produced from over 10% of the corn crop. Ethanol demand is expected to more than double in the next ten years. For the supply to be available to meet this demand, new technologies must be moved from the laboratories to commercial reality. The world ethanol production is about 60% from feedstock from sugar crops.

The corn starch-to-fuel ethanol industry has grown during the past 30 years by bioethanol researchers. The majority of bioethanol researchers are focusing on the challenge of producing bioethanol from lignocellulosic biomass instead of corn starch. Toward this end, the researchers already have developed effective technol-

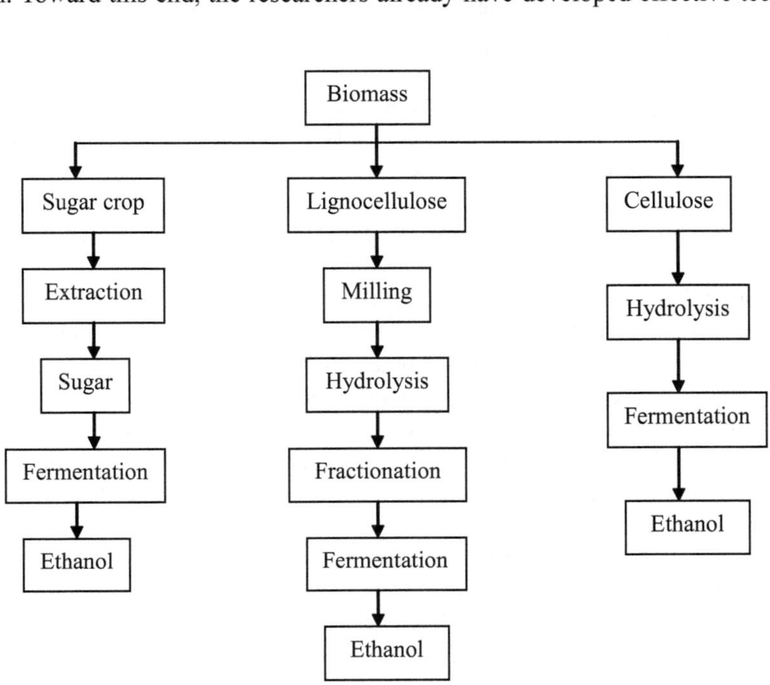

Fig. 7.1 Ethanol production from sugar crops and lignocellulosic feedstocks

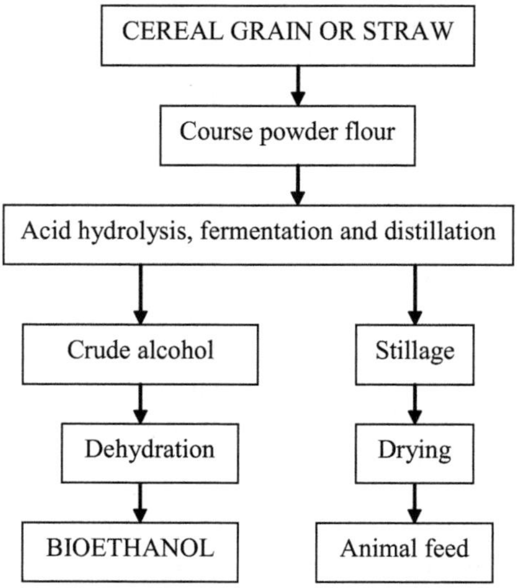

Fig. 7.2 Flow chart for the production bioethanol from cereal grain or straw

ogy to thermochemically pretreat biomass: to hydrolyze hemicellulose to break it down into its component sugars and open up the cellulose to treatment, to enzymatically hydrolyze cellulose to break it down to sugars, and to ferment both five-carbon sugars from hemicellulose and six-carbon sugars from cellulose.

Cellulose is a remarkable pure organic polymer, consisting solely of units of anhydroglucose held together in a giant straight-chain molecule. Cellulose must be hydrolyzed to glucose before fermentation to ethanol. Conversion efficiencies of cellulose to glucose may be dependent on the extent of chemical and mechanical pretreatments to structurally and chemically alter the pulp and paper mill wastes. The method of pulping, the type of wood, and the use of recycled pulp and paper products also could influence the accessibility of cellulose to cellulase enzymes. Figure 7.2 shows the flow chart for the production of bioethanol from cereal grain or straw.

Cellulose fraction of the structural components is insoluble in most solvents and has a low accessibility to acid and enzymatic hydrolysis. Hemicelluloses (arabinoglycuronoxylan and galactoglucomannans) are related to plant gums in composition, and occur in much shorter molecule chains than cellulose. The hemicelluloses, which are present in deciduous woods chiefly as pentosans and in coniferous woods almost entirely as hexosanes, undergo thermal decomposition very readily. Hemicelluloses are derived mainly from chains of pentose sugars, and act as the cement material holding together the cellulose micelles and fiber. Hemicelluloses are largely soluble in alkali and, as such, are more easily hydrolyzed.

Bioethanol can be produced from a large variety of carbohydrates with a general formula of $(CH_2O)_n$. The chemical reaction is composed of enzymatic hydrolysis of sucrose followed by fermentation of simple sugars. Fermentation of sucrose is performed using commercial yeast such as *Saccharomyces cerevisiae*. First, invertase enzyme in the yeast catalyzes the hydrolysis of sucrose to convert it into glucose and fructose.

$$C_{12}H_{22}O_{11} \rightarrow C_6H_{12}O_6 + C_6H_{12}O_6 \quad (7.1)$$

Sucrose Glucose Fructose

Second, zymase, another enzyme also present in the yeast, converts the glucose and the fructose into ethanol.

$$C_6H_{12}O_6 \rightarrow 2C_2H_5OH + 2CO_2 \quad (7.2)$$

Glucoamylase enzyme converts the starch into D-glucose. The enzymatic hydrolysis is then followed by fermentation, distillation and dehydration to yield anhydrous bioethanol. Corn (60–70% starch) is the dominant feedstock in the starch-to-bioethanol industry worldwide.

The lignocellulose is subjected to delignification, steam explosion and dilute acid prehydrolysis, which is followed by enzymatic hydrolysis and fermentation into bioethanol (Sokhansanj et al. 2002; Kim and Dale 2005). A major processing step in an ethanol plant is enzymatic saccharification of cellulose to sugars through treatment by enzymes; this step requires lengthy processing and normally follows a short-term pretreatment step (Kumar et al. 2005).

Hydrolysis breaks down the hydrogen bonds in the hemicellulose and cellulose fractions into their sugar components: pentoses and hexoses. These sugars can then be fermented into bioethanol. The most commonly applied methods can be classified in two groups: chemical hydrolysis (dilute and concentrated acid hydrolysis) and enzymatic hydrolysis. In the chemical hydrolysis, the pretreatment and the hydrolysis may be carried out in a single step. There are two basic types of acid hydrolysis processes commonly used: dilute acid and concentrated acid.

The biggest advantage of dilute acid processes is their fast rate of reaction, which facilitates continuous processing. Since five-carbon sugars degrade more rapidly than six-carbon sugars, one way to decrease sugar degradation is to have a two-stage process. The first stage is conducted under mild process conditions to recover the five-carbon sugars while the second stage is conducted under harsher conditions to recover the six-carbon sugars.

Concentrated sulfuric or hydrochloric acids are used for hydrolysis of lignocellulosic materials. The concentrated acid process uses relatively mild temperatures, and the only pressures involved are those created by pumping materials from vessel to vessel. Reaction times are typically much longer than for dilute acid. This process provides a complete and rapid conversion of cellulose to glucose and hemicelluloses to five-carbon sugars with little degradation. The critical factors needed to make this process economically viable are to optimize sugar recovery and cost effectively recovers the acid for recycling. The solid residue from the first stage is dewatered and soaked in a 30 to 40% concentration of sulfuric acid for 1 to 4 h as a precellulose hydrolysis step. The solution is again dewatered and dried, increasing the acid concentration to about 70%. After reacting in another vessel for 1 to 4 h at low temperatures, the contents are separated to recover the sugar and acid. The sugar/acid solution from the second stage is recycled to the first stage to provide the acid for the first stage hydrolysis (Demirbas 2006). The primary advantage of the concentrated acid process is the potential for high sugar recovery efficiency. The acid and sugar are separated via ion exchange and then acid is reconcentrated via multiple effect evaporators.

7.4 Production of Chemicals

In recent years, extraordinary progress has been made in the development of practical processes and products from biopolymers such as starch, cellulose, and lactic acid. Starch is an inexpensive, annually renewable material derived from corn and other crops. Cellulose, with its linear molecules, is in general more suitable for material processing into fibers and films. The degrees of polymerization of celluloses of different origin are between 2500 and 8000. Selective conversion of cellulose to small organic molecules, including carboxylic acids, represents a potential route for upgrading biomass resources to value-added chemical precursors. However, since biomass resources such as cellulose are not soluble in water, the use of a catalyzed system requires a novel reactor design that facilitates slurry flow (Schutt et al. 2002). Lignocellulosic materials from agriculture and forest management are the largest sources of hexose (six-carbon) and pentose (five-carbon) sugars with a potential for the production of biofuels, chemicals and other economic byproducts (Koutinas et al. 2007).

In general, biodegradable plastics whose components are derived from renewable raw materials can be made from abundant agricultural/animal resources like cellulose, starch, collagen, casein, just like soy protein polyesters and triglycerides. Use of plastics in agriculture and rural communities has extensively increased (Orhan et al. 2004). The increasing use of polymers for disposable items such as packaging creates problems for garbage disposal and has produced a demand for biodegradable polymers (Rutkowska et al. 2000). Biodegradability of plastics has been proposed as a solution for the waste problem. Biodegradable plastics have an expanding range of potential applications and they are environmentally friendly, since they degrade more rapidly than conventional disposables. Biodegradable plastics degrade over a period of time if exposed to sunlight and air. This biodegradation is caused by biological activity, particularly by enzyme action leading to significant changes in the material's chemical structure. A cereal-based biorefinery for the production of biodegradable plastics is depicted in Fig. 7.3 (Koutinas et al. 2007).

Sugar crops like sugar beet, or sugar cane (sucrose) or corn (starch), offer specific major components for certain products, and they are compelling reasons for establishing a biorefinery system. During the fermentation process sugars are converted into lactic acid and proteins are hydrolyzed to free amino acids (Thang and Novalin 2008). Therefore, lactic acid and amino acids are seen as the key compounds in a "green biorefinery" system based on grass silage (Kromus et al. 2004). Besides amino acids, the interest in producing lactic acid is growing because of its great potential in the manufacturing of biodegradable polymers (Datta et al. 1995). However, one main difficulty in lactic acid production lies in the recovery cost.

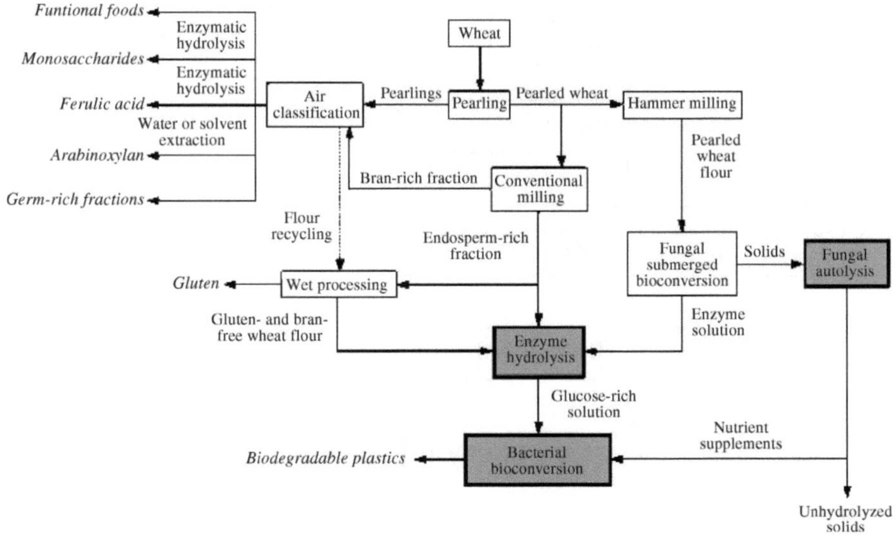

Fig. 7.3 A wheat-based biorefinery for the production of biodegradable plastics

Methanol is currently made from natural gas but can also be made using biomass via partial oxidation reactions (Demirbas 2008). Biomass and coal can be considered as a potential fuel for gasification and further syngas production and methanol synthesis (Takezewa et al. 1987). Adding sufficient hydrogen to the synthesis gas to convert all of the biomass into methanol carbon would more than double the methanol produced from the same biomass base (Phillips et al. 1990). Waste material can be partially converted to methanol, where the product yield for the conversion process is estimated to be 185 kg of methanol per metric ton of solid waste (Brown et al. 1952; Sorensen 1983). Agriculture (m)ethanol is at present more expensive than synthesis ethanol from ethylene and methanol from natural gas (Grassi 1999).

Before modern production technologies were developed in the 1920s, methanol was obtained from wood as a co-product of charcoal production and, for this reason, was commonly known as wood alcohol. However, the yield from this method of production was very low. One ton of hardwood would only yield 1 or 2%. This led to its eventual replacement by less expensive alternatives. Biomass resources can be used to produce methanol. The pyroligneous acid obtained from wood pyrolysis consists of about 50% methanol, acetone, phenols and water. Biomass resources include crop residues, forage, grass, crops, wood resources, forest residues, short-rotation wood energy crops and the lignocellulosic components of municipal solid waste. As a renewable resource, biomass represents a potentially inexhaustible supply of feedstock for methanol production. Methanol is currently made from natural gas but can be made using wood waste or garbage via partial oxidation reaction into syngas, followed by the catalytic conversion into methanol, known as biomethanol. Current natural gas feedstocks are so inexpensive that even with tax incentives renewable methanol has not been able to compete economically. Technologies are being developed that may eventually result in commercial viability of renewable methanol.

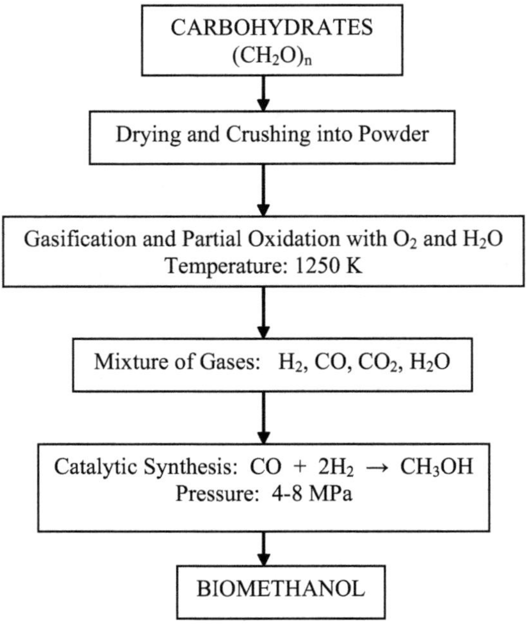

Fig. 7.4 Biomethanol from carbohydrates by gasification and partial oxidation with O_2

Biomass can be considered a potential fuel for gasification and further syngas production and methanol synthesis. The feasibility of achieving the conversion has been demonstrated in a large-scale system in which a product gas is initially produced by pyrolysis and gasification of a carbonaceous matter. The synthetic process developed in 1927 replaced this traditional method. Methanol from biochar was about 1–2% of volume or 6 gallons per ton. Syngas from biomass is altered by catalyst under high pressure and temperature to form methanol. This method will produce 100 gallons of methanol per ton of feed material. Figure 7.4 shows biomethanol from carbohydrates by gasification and partial oxidation with O_2 and H_2O.

7.4.1 Fungal Enzymes to Convert Biomass to Bioethanol

Bioethanol production from biomass is one such already proven industrial process for renewable energy production. Bioethanol can be directly mixed with petrol and used in today's cars, or converted to electricity. Currently there are two types of blends of ethanol/gasoline on the market: blends of 10% ethanol and 90% gasoline (E10) and blends of 85% ethanol and 15% gasoline (E85). In the US, many states currently mandate E10 in gasoline. It is generally accepted that bioethanol gives a 70% carbon dioxide reduction, which means 7% in an E10 blend or 50% in an E85 blend. Increased bioethanol usage could reduce US greenhouse gas emissions to 1.7 billion tons per year. Recent research has established that polysaccharides in biomass can be hydrolyzed enzymatically into glucose sugar that is fermented to bioethanol.

When manufactured from agricultural sources, like corn or wood, ethanol is commonly referred to as bioethanol. Theoretically, bioethanol production should be able to yield 0.5 g of ethanol per gram of raw biomass, which translates into an energy recovery of approximately 90%. Corn is a common substrate for bioethanol manufacture because the process is relatively free of technical obstacles. Microorganisms are involved in the transformation of corn to bioethanol in two ways: they catalyze the hydrolysis of starches using amylases and amyloglucosidases, and they ferment the resulting sugars to bioethanol. The fermentation step is generally carried out by yeast, but certain strains of bacteria, including *Zymomonas mobilis* and recombinant strains of *Escherichia coli* and *Klebsiella oxytoca*, are also capable of producing high yields of bioethanol.

Because bioethanol is a liquid, bioethanol fits into the current fuel infrastructure, although transport requires special handling to prevent water accumulation. Bioethanol can be cost-competitive with petroleum. However, using solid substrates in converting lignocellulose to sugars poses a dilemma in bioethanol production. Cellulose and lignocellulose are in much greater supply than starch and sugars and are, therefore, preferred substrates for ethanol production. However, producing ethanol from cellulose and lignocellulose is comparatively difficult and expensive.

Lignocellulose includes such diverse sources as switchgrass, corn stalks, and wood chips. Fungal enzymes and fermentative yeasts are then used to transform lignocellulose first to sugars and then to bioethanol. The low lignocellulose reactivity limits the production of bioethanol. The expense of enzyme production is the biggest economic barrier to lignocellulose conversion to bioethanol.

Biological pretreatments use fungi to solubilize the lignin. Biodelignification is the biological degradation of lignin by microorganisms. This process was first mentioned in 1984 as possibly useful in the future, although at that time it was inadequate and expensive, required a long process time and the microorganisms were poisoned by lignin derivatives. These technologies could greatly simplify pretreatment, but the rates are slow, yields are low, and little experience has been developed with such approaches.

Recent research has focused on enzyme catalysts called *cellulases* that can attack these chains more efficiently, leading to very high yields of fermentable sugars. Fungal *cellulases* and *beta-glucosidases* produced in separate aerobic reactors can be extracted in very high yields, but because these enzymes have low specific activities, they must be used in large quantities to achieve lignocellulose conversion.

Fungi produce a plethora of enzymes that are used to degrade complex polysaccharides and proteins into simpler sugars and amino acids, and are long established as a key source of a wide variety of industrially important enzymes. Some of these enzymes have already been harnessed in releasing fermentable sugars from a variety of biomass feedstocks, including waste paper, food stuffs, cereals, sugar crops, grains, and woods.

Both bacteria and fungi can produce cellulases for the hydrolysis of lignocellulosic materials. These microorganisms can be aerobic or anaerobic, mesophilic or thermophilic. Bacteria belonging to *Clostridium*, *Cellulomonas*, *Bacillus*, *Thermomonospora*, *Ruminococcus*, *Bacteriodes*, *Erwinia*, *Acetovibrio*, *Microbispora*, and *Streptomyces* can produce cellulases (Sun 2002).

The widely accepted mechanism for enzymatic cellulose hydrolysis involves synergistic actions by endoglucanases or endo-1,4-β-glucanases (EG), exoglucanases or cellobiohydrolases (CBH), and β-glucosidases (BGL). EG play an important role in the cellulose hydrolysis by cleaving cellulose chains randomly and thus encouraging strong degradation. EG hydrolyze accessible intramolecular β-1,4-glucosidic bonds of cellulose chains randomly to produce new chain ends, exoglucanases cleave cellulose chains at the ends to release soluble cellobiose or glucose, and BGL hydrolyze cellobiose to glucose in order to eliminate cellobiose inhibition (Zhang et al. 2006). BGL complete the hydrolysis process by catalyzing the hydrolysis of cellobiose to glucose.

Filamentous fungi are the major source of *cellulases* and *hemicellulases*. Mutant strains of *Trichoderma sp.* (*T. viride, T. reesei, T. longibrachiatum*) have long been considered to be the most productive and powerful destroyers of crystalline cellulose (Gusakov et al. 2007). CBH I and CBH II are the major *T. reesei* enzymes, the content of CBH I comprises up to 60% of the total cellulolytic protein, whereas, the content of CBH II is about 20% (Gusakov et al. 2005). Similarly, EG I and EG II are the dominant endoglucanases in *T. reesei*, and presumably act as important partners to CBH I in nature (Väljamäe et al. 2001). Such protein yields are comparable or exceed the respective parameters for the best *Trichoderma sp.* strains (35–40 g/L) (Gusakov et al. 2007). Yeast and fungi tolerate a range of 3.5 to 5.0 pH. The ability to lower pH below 4.0 offers a method for present operators using yeast in less than aseptic equipment to minimize loss due to bacterial contaminants.

Fungal lignocellulolytic enzymes for conversion of lignocellulosic biomass to fermentable sugars for the production of bioethanol were used (Tabka et al. 2006). Wheat straw was pretreated by acid treatment with diluted sulfuric acid followed by steam explosion. Several enzymatic treatments implementing hydrolases (cellulases and xylanases from *Trichoderma reesei*, recombinant feruloyl esterase (FAE) from *Aspergillus niger*) and oxidoreductases (laccases from *Pycnoporus cinnabarinus*) were investigated for the saccharification of exploded wheat straw. A synergistic effect between cellulases, FAE and xylanase was proven under a critical enzymatic concentration (10 U/g of cellulases, 3 U/g of xylanase, and 10 U/g of FAE). The yield of enzymatic hydrolysis was enhanced by increasing the temperature from 110 to 323 K, and by the addition of a non-ionic surfactant, Tween 20 (Tabka et al. 2006).

7.5 Upgrading of Biochemical Conversion Products

Hydrolysis is often used to pretreat lignocellulosic feedstocks to break down the cellulose and hemicelluloses from the lignocellulose and break down the compounds into simple sugars. Hydrolysis can be catalyzed by use of acids (either strong or weak), enzymes, and/or hydrothermal means, the latter including hot water and supercritical methods. Figure 7.5 shows the flow diagram of the enzymatic hydrolysis process.

Fermentation is generally used industrially to convert substrates such as glucose to ethanol for use in beverage, fuel, and chemical applications, and to other chem-

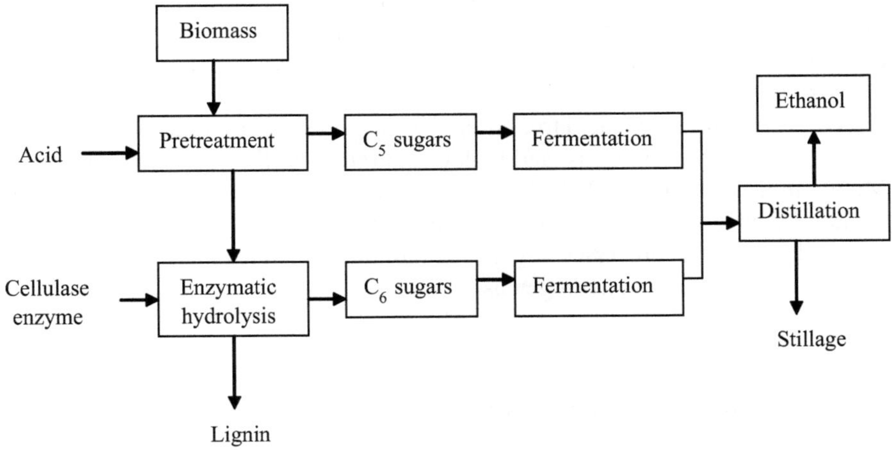

Fig. 7.5 Upgrading of biomass via enzymatic hydrolysis process

icals (e.g., lactic acid used in producing renewable plastics) and products (e.g., enzymes for detergents). Strictly speaking, fermentation is an enzymatically controlled anaerobic process although the term is sometimes more loosely applied to include aerobic processing as well.

As a sustainable energy source, hydrogen is a promising alternative to fossil fuels. It is a clean and environmentally friendly fuel, due to its high conversion efficiency, recyclability and non-polluting nature (Han and Shin 2004).

Hydrogen produced from water, biorenewable feedstocks, either biologically (biophotolysis and fermentation) or photobiologically (photodecomposition), is known as biohydrogen. Biohydrogen technology will play a major role in the future

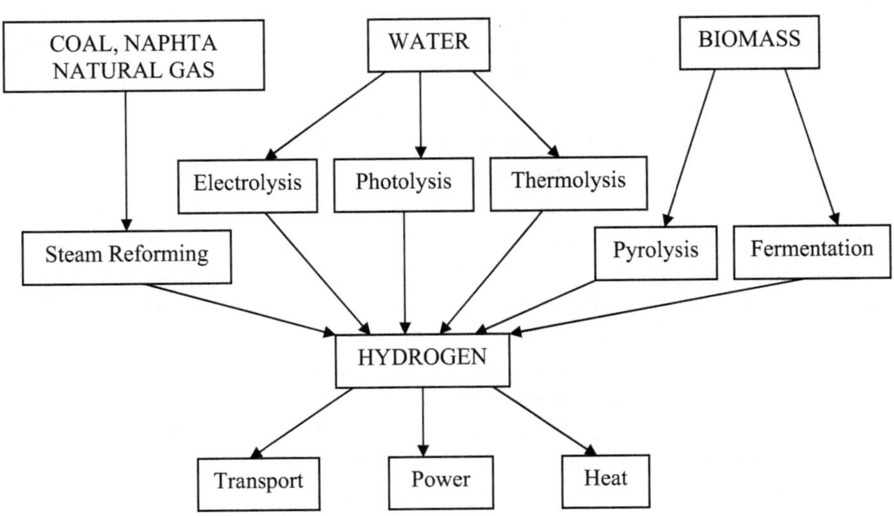

Fig. 7.6 Main alternative processes of hydrogen production and hydrogen use

because it can utilize renewable sources of energy. Hydrogen is currently more expensive than conventional energy sources, which limits its use, but researchers are looking into different technologies to produce hydrogen economically from biomass (Nath and Das 2003).

Hydrogen can be generated from water by electrolysis, photolysis, direct thermal decomposition or thermolysis, and biological processes (Das and Veziroglu 2001; Momirlan and Veziroglu 2002). Many studies have reported on biohydrogen production by photocatalytic (Hwang et al. 2004) or enzymatic (de Vrije et al. 2002; Han and Shin 2004) processes. Figure 7.6 shows the main alternative processes of hydrogen production and hydrogen use.

7.6 Refining of Biochemical Conversion Products

Biorefineries for production of several products and byproducts such as biofuels, heat and/or electricity have been in focus in the recent years (Zhang 2008). In a biorefinery, biomass can be converted to useful biomaterials and/or energy carriers in an integrated manner; this then maximizes the economic value of the biomass used, while reducing the waste streams produced (Koutinas et al. 2007).

The biorefinery system starts with the contract harvesting of whole crops (grain and straw), which are then stored and fractionated (including drying as necessary) into products and byproducts for sale. A biorefinery is a factory that processes crops, such as wheat, barley and oilseed rape, to produce various refined specialized fractions, such as flour, gluten, starch, oil, straw chips, etc. The concept of a biorefinery, compared with for example a flour mill, is that the use and value of all the fractions into which the input can be separated is maximized (Audsley and Annetts 2003).

The analysis of a biorefinery system can be considered in three parts. The first is the effect on the farm of selling products to a biorefinery, on the assumption that the biorefinery contracts to harvest the crop using a wholecrop forage harvester. The second part is the impact of the type of biorefinery system on the profitability of the processing required to produce the various products. The third part is the transport of crop to the biorefinery, which is a function of the distribution of farms around the biorefinery location (Audsley and Annetts 2003).

As biomass hydrolysis, and sugar fermentation technologies approach commercial viability, advancements in product recovery technologies will be required. For cases in which fermentation products are more volatile than water, recovery by distillation is often the technology of choice. Distillation technologies that will allow the economic recovery of dilute volatile products from streams containing a variety of impurities have been developed and commercially demonstrated. A distillation system separates the bioethanol from water in the liquid mixture.

The first step is to recover the bioethanol in a distillation or beer column, where most of the water remains with the solid part. The product (37% bioethanol) is then concentrated in a rectifying column to a concentration just below the azeotrope (95%). The remaining bottom product is fed to the stripping column to remove

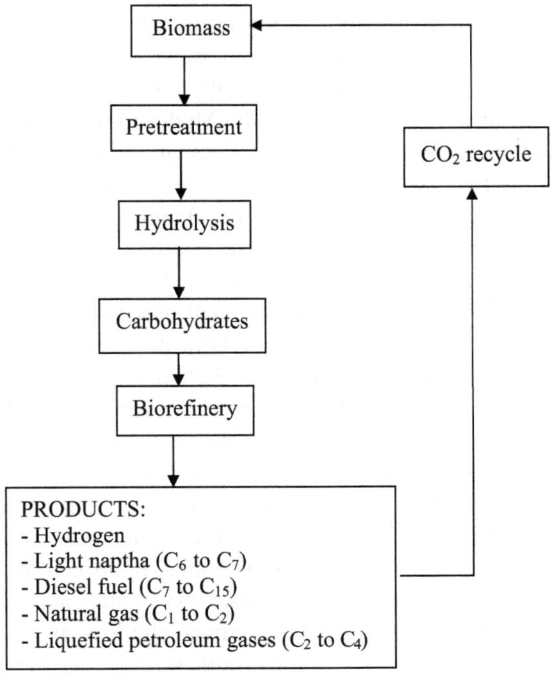

Fig. 7.7 An integrated biorefinery for conversion of carbohydrates to fuels by aqueous-phase processing

additional water, with the bioethanol distillate from stripping being recombined with the feed to the rectifier. The recovery of bioethanol in the distillation columns in the plant is fixed to be 99.6% to reduce bioethanol losses.

After the first effect, solids are separated using a centrifuge and dried in a rotary dryer. A portion (25%) of the centrifuge effluent is recycled to fermentation and the rest is sent to the second and third evaporator effects. Most of the evaporator condensate is returned to the process as fairly clean condensate (a small portion, 10%, is split off to wastewater treatment to prevent buildup of low-boiling compounds) and the concentrated syrup contains 15–20% by weight total solids.

One of the advantages of alkane production from biomass by aqueous-phase dehydration/ hydrogenation is that the alkanes spontaneously separate from the aqueous feed solution, whereas ethanol produced during fermentation processes must be removed from water by an energy-intensive distillation step.

Biomass-derived oxygenates can be converted into hydrogen and alkanes (ranging from C_1 to C_{15}) via aqueous-phase processing (Audsley and Annetts 2003). These aqueous phase processes could be used in an integrated biorefinery to produce a range of fuels, as shown in Fig. 7.7. The first step in the biorefining process is conversion of biomass into an aqueous sugar solution. Production of hydrogen for biorefining processes is accomplished by aqueous-phase reforming. The biorefinery can also produce light alkanes ranging from C_1 to C_6 by aqueous-phase dehydration/hydrogenation (Audsley and Annetts 2003). The light alkanes could be used as synthetic natural gas, liquefied petroleum gas, and a light naptha stream. Aqueous-phase processing can also produce larger alkanes ranging from C_7 to C_{15} by combining the dehydration/hydrogenation reactions with an aldol condensation step prior to the aqueous-phase dehydration/hydrogenation step (Huber et al. 2005).

Biomass has been traditionally converted to liquid fuels by either (a) fermentation or (b) pyrolysis methods. Modern improvements to these classical processes are many in number but do not essentially change the type of product resulting from these two vastly different sets of reaction conditions. While ethanol production by fermentation has become more efficient, it is still limited to a 67% yield due to the loss of one-third of the available carbon as carbon dioxide gas. Pyrolytic reactions also lose carbon as gases and char but may achieve about 80% carbon conversion. While most thermochemical processes usually require nearly dry feedstock, the hydrothermal upgrading (HTU) process requires a 3:l ratio of water to biomass. However, HTU produces only 50% biocrude, which still contains 10–15% oxygen. Obviously, there remains a need for a variety of fuels from many sources, especially conventional liquid fuels for transportation purposes. To resolve this fuel problem and to use a renewable resource, a strategy was selected to prepare valuable liquid hydrocarbons from biomass by a new chemical process.

Glycerol can be converted into higher-value products, such as 1,3-propanediol, 1,2-propanediol, dihydroxyacetones, hydrogen, polyglycerols, succinic acid, and polyesters. Main glycerol-based oxygenates are 1,3-propanediol, 1,2-propanediol, propanol, glycerol tert-butyl ethers, ethylene glycol and propylene glycol. Glycerol has been pyrolyzed for the production of clean fuels such as H_2 or a feedstock such as syngas for additional transportation fuel via Fischer–Tropsch synthesis (FTS). The conversion of glycerol to H_2 and CO takes place according to the following stoichiometric equation:

$$C_3O_3H_8 \rightarrow 3CO + 4H_2 \tag{7.3}$$

The stoichiometry for conversion of glycerol to liquid alkanes, by the formation of synthesis gas coupled with FTS, is shown in Eq. 7.4:

$$25C_3O_3H_8 \rightarrow 7C_8H_{18} + 19CO_2 + 37H_2O \tag{7.4}$$

This overall reaction to produce liquid fuels from glycerol is slightly exothermic and the yield of liquid alkanes is 40% at 1.7 MPa pressure.

It is possible to produce light alkanes by aqueous-phase reforming of biomass-derived oxygenates such as sorbitol, which can be obtained from glucose by hydrogenation (Huber et al. 2005; Metzger 2006). The production of alkanes from aqueous carbohydrate solutions would be advantageous because of the easy separation of the alkanes from water. Much hydrogen is needed to reduce biomass-derived oxygenates to alkanes as shown in Eq. 7.5:

$$C_6H_{14}O_6 + 6H_2 \rightarrow C_6H_{14} + 6H_2O \tag{7.5}$$

It was shown in earlier work that hydrogen can be produced in a catalytic process at 500 K in water from biomass-derived polyols having a C/O stoichiometry of 1:1 such as sorbitol, as shown in Eq. 7.6 (Metzger 2006):

$$C_6H_{14}O_6 + 6H_2O \rightarrow 13H_2 + 6CO_2 \tag{7.6}$$

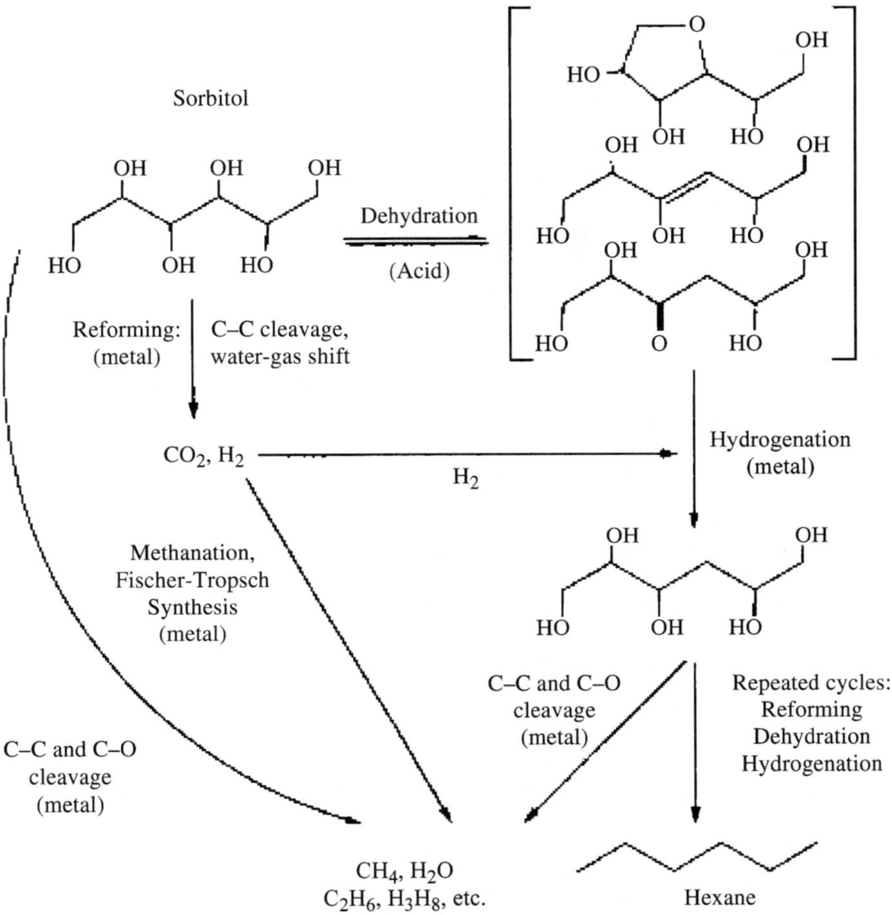

Fig. 7.8 Reaction pathways for the production of alkanes from sorbitol in aqueous solution over a $Pt/SiO_2/Al_2O_3$ catalyst

The overall reaction (Eq. 7.7) is exothermic:

$$19C_6O_6H_{14} \rightarrow 13C_6H_{14} + 36CO_2 + 42H_2O \tag{7.7}$$

This process converts approximately 1.5 mol of sorbitol into 1 mol of hexane. About 95% of the heating value but only 30% of the mass of the biomass-derived reactant is retained in the product. The remaining 70% of the biomass is found as CO_2 and water. A maximum yield of 61% of hydrogen with respect to sorbitol was obtained over Pt/Al_2O_3 under optimized reaction conditions. A rationalization of the reaction pathway is given in Fig. 7.8 (Metzger 2006).

References

Audsley, E., Annetts, J.E. 2003. Modelling the value of a rural biorefinery, part I: the model description. Agric Sys 76:39–59.

Bala, B.K. 2005. Studies on biodiesels from transformation of vegetable oils for diesel engines. Energy Edu Sci Technol 15:1–45.

Brown, H.P., Panshin, A.J., Forsaith, C.C. 1952. Textbook of wood technology, vol. II. McGraw-Hill, New York.

Das, D., Veziroglu, T.N. 2001. Hydrogen production by biological processes: a survey of literature. Int J Hydrogen Energy 26:13–28.

de Vrije, T., de Haas, G.G., Tan, G.B., Keijsers, E.R.P., Claassen, P.A.M. 2002. Pretreatment of Miscanthus for hydrogen production by *Thermotoga elfii*. Int J Hydrogen Energy 27:1381–1390.

Datta, R., Tsai, S.P., Bonsignore, P., Moon, S.H., Frank, J.R., 1995. Technological and economic potential of poly (lactic acid) and lactic acid derivatives. FEMS Microbiol Rev 16:221–231.

Demirbas, A. 2005. Bioethanol from cellulosic materials: a renewable motor fuel from biomass. Energy Sources 27:327–337.

Demirbas, A. 2006. Global biofuel strategies. Energy Edu Sci Technol 17:27–63.

Demirbas, A. 2008a. Biofuels sources, biofuel policy, biofuel economy and global biofuel projections. Energy Convers Manage 49:2106–2116.

Demirbas, A. 2008b. Economic and environmental impacts of the liquid biofuels. Energy Edu Sci Technol 22:37–58.

Demirbas, A. Biodegradability of biodiesel and petrodiesel fuels. Energy Sources Part A 3:169–174.

Du, W., Xu, Y., Liu, D., Zeng, J. 2004. Comparative study on lipase-catalyzed transformation of soybean oil for biodiesel production with different acyl acceptors. J Molecular Catal B Enzym 30:125–129.

Grassi, G. 1999. Modern bioenergy in the European Union. Renewable Energy 16(1–4):985–990.

Gusakov, A.V., Salanovich, T.N., Antonov, A.I., Ustinov, B.B., Okunev, O.N., Burlingame, R., Emalfarb, M., Baez, M., Sinitsyn, A.P. 2007. Design of highly efficient cellulase mixtures for enzymatic hydrolysis of cellulose. Biotechnol Bioeng 97:1028–1038.

Gusakov, A.V., Sinitsyn, A.P., Salanovich, T.N., Bukhtojarov, F.E., Markov, A.V., Ustinov, B.B., van Zeijl, C., Punt, P., Burlingame, R. 2005. Purification, cloning and characterisation of two forms of thermostable and highly active cellobiohydrolase I (Cel7A) produced by the industrial strain of *Chrysosporium lucknowense*. Enzyme Microbial Technol 36:57–69.

Han, S.-K., Shin, H.-S. 2004. Biohydrogen production by anaerobic fermentation of food waste. Int J Hydrogen Energy 29:569–577.

Huber, G.W., Chheda, J.N, Barrett, J.A. 2005. Dumesic, production of liquid alkanes by aqueous phase processing of biomass-derived carbohydrates. Science 308:1446–1450.

Hwang, D.W., Kim, H.G., Jang, J.S., Bae, S.W., Ji, S.M., Lee, J.S. 2004. Photocatalytic decomposition of water-methanol solution over metal-doped layered perovskites under visible light irradiation. Catal Today 93–5:845–850.

Kim, S., Dale, B.E. 2005. Life cycle assessment of various cropping systems utilized for producing biofuels: Bioethanol and biodiesel. Biomass Bioenergy 29:426–439.

Koutinas, A.A., Xu, Y., Wang. R., Webb, C. 2007. Polyhydroxybutyrate production from a novel feedstock derived from a wheat-based biorefinery. Enzyme Microbial Technol 40:1035–1044.

Kromus, S., Wachter, B., Koschuh, W., Mandl, M., Krotscheck, C., Narodoslawsky, M., 2004. The Green Biorefinery Austria: development of an integrated system for green biomass utilization. Chem Biochem Eng Quart 18:7–12.

Kumar, A., Cameron, J.B., Flynn, P.C. 2005. Pipeline transport and simultaneous saccharification of corn stover. Biores Technol 96:819–829.

Metzger, J.O. 2006. Production of liquid hydrocarbons from biomass. Angew Chem Int Ed 45:696–698.

Momirlan, M., Veziroglu, T. 2002. Current status of hydrogen energy. Renew Sustain Energy Rev 6:141–179.

Nath, K., Das, D. 2003. Hydrogen from biomass. Current Sci 85:265–271.

Noureddini, H., Gao, X., Philkana, R.S. 2005. Immobilized Pseudomonas cepacia lipase for biodiesels fuel production from soybean oil. Biores Technol 96:769–777.

Orhan, Y., Hrenovic, J., Buyukgungor, H. 2004. Biodegradation of plastic compost bags under controlled soil conditions. Acta Chim Slov 51:579–588.

Phillips, V.D., Kinoshita, C.M., Neill, D.R., Takashi, P.K. 1990. Thermochemical production of methanol from biomass in Hawaii. Appl Energy 35:167–175.

Rutkowska, M., Krasowska, K., Heimowska, A., Smiechowska, M., Janik, H. 2000. The influence of different processing additives on biodegradation of polyepsilon-caprolactone. Iranian Polymer J 9:221–229.

Schutt, B.D., Serrano, B., Cerro, R.L., Abraham, M.A. 2002. Production of chemicals from cellulose and biomass-derived compounds through catalytic sub-critical water oxidation in a monolith reactor. Biomass Bioenergy 22:365–375.

Shieh, C.-J., Liao, H.-F., Lee, C.-C. 2003. Optimization of lipase-catalyzed biodiesel by response surface methodology. Biores Technol 88:103–106.

Sokhansanj, S., Turhollow, A., Cushman, J., Cundiff, J. 2002. Engineering aspects of collecting corn stover for bioenergy. Biomass Bioenergy 23:347–355.

Sorensen, H.A. 1983. Energy conversion systems. Wiley, New York.

Sun, Y. 2002. Enzymatic hydrolysis of rye straw and Bermuda grass for ethanol production. Dissertation, Biological and Agricultural Engineering, North Carolina State University, August 12, 2002.

Tabka, M.G., Herpoël-Gimbert, I., Monod, F., Asther, M., Sigoillot, J.C. 2006. Enzymatic saccharification of wheat straw for bioethanol production by a combined cellulase xylanase and feruloyl esterase treatment. Enzyme Microbial Technol 39:897–902.

Takezawa, N., Shimokawabe, M., Hiramatsu, H., Sugiura, H., Asakawa, T., Kobayashi, H. 1987. Steam reforming of methanol over Cu/ZrO_2. Role of ZrO_2 support. React Kinet Catal Lett 33:191–196.

Thang, V.H., Novalin, S. 2008. Green biorefinery: separation of lactic acid from grass silage juice by chromatography using neutral polymeric resin. Biores Technol 99:4368–4379.

Väljamäe, P., Pettersson, G., Johansson, G. 2001. Mechanism of substrate inhibition in cellulose synergistic degradation. Eur J Biochem 268:4520–4526.

Zhang, Y.H.P., Himmel, M.E., Mielenz, J.R. 2006. Outlook for cellulase improvement: screening and selection strategies. Biotechnol Adv 24:452–481.

Zhang, Y. 2008. Reviving the carbohydrate economy via multi-product lignocellulose biorefineries. J Ind Microbiol Biotechnol 35:367–375.

Chapter 8
Biorefining Economy

8.1 Introduction

A biorefinery system includes the harvesting, storage and transport of the products as well as the biorefinery itself. Advanced biorefineries are envisioned to serve as the foundation of a new bioindustry. With the rise in oil and natural gas prices, bio-based feedstocks are becoming increasingly competitive. The biorefinery of the future is likely to integrate both bioconversion and thermochemical conversion technologies.

The biorefinery economy is a vision for a future in which biorenewables replaces fossil fuels. The transition to a biorefinery economy would require huge investment in new infrastructure to produce, store and deliver biorefinery products to end users. The transition to the biorefinery economy could take several decades, because of the slow turnover of the existing stock of capital. The transition to a biorefinery economy is likely to begin later in most developing economies than in the industrialized countries. The earlier industrialized countries/economies begin the transition to biorefinery economy, the quicker they could achieve energy sustainability (UNEP 2006).

Biomass, in particular, could be a low-cost option for some countries. Therefore, a cost-effective energy-production process could be achieved in which agricultural wastes and various other biomasses are recycled to produce biorefinery products economically.

Estimations for the cost of biorefinery products are affected by a range of drivers that could change in direction and importance over time. These include:

1. Supply cost, market price and demand
2. Competing, non-energy markets for biomass
3. Preferences of farmers and woodland owners
4. Success of alternative waste recovery and recycling
5. Production cost

6. Storage cost
7. Distribution cost
8. Access to market

Biomass residues are the cheapest feedstocks. Because of the low sulfur content of biomass, a sulfur removal system is not likely to be required. Currently, biomass-driven combined heat and power, co-firing, and combustion plants provide reliable, efficient, and clean power and heat. Production and use of biofuels are growing at a very rapid pace. Sugar cane-based ethanol is already a competitive biofuel in tropical regions. In the medium term, ethanol and high-quality synthetic fuels from woody biomass are expected to be competitive at crude oil prices above US $45 per barrel. Figure 8.1 shows the value-added wood products. The pyrolysis-based technology, in particular, because of the co-products opportunity, has the most favorable economics. With scientific and engineering advancements, biomass can be viewed as a key and economically viable component to a renewable-based products economy.

The cost of large-scale production of bio-based products is currently high in developed countries. For example, the production cost of biofuels may be three times higher than that of petroleum fuels, without, however, considering the non-

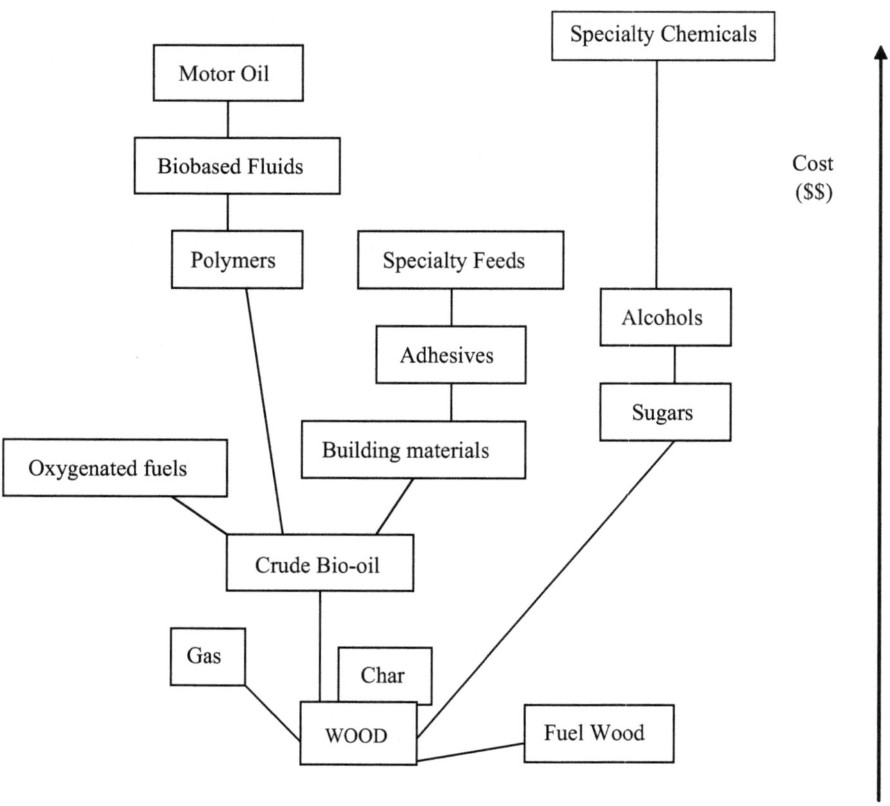

Fig. 8.1 Value-added wood products (cost increases from *bottom* to *top*)

Table 8.1 Some technical properties of fuels

Fuel property	Gasoline	Diesel No. 2	Iso-octane	Methanol	Ethanol
Cetane number	–	50	–	5	8
Octane number	96	–	100	114	108
Auto-ignition temperature (K)	644	588	530	737	606
Lower heating value (MJ/kg)	44	43	45	20	27

market benefits. Conversely, in developing countries, the costs of producing biofuels are much lower than in the OECD countries and very near to the world market price of petroleum fuel (UN 2006).

It is well known that transport is almost totally dependent on fossil fuels, particularly petroleum-based fuels such as gasoline, diesel fuel, liquefied petroleum gas, and compressed natural gas. Some technical properties of fuels are given in Table 8.1. As the amount of available petroleum decreases, the need increases for alternate technologies to produce liquid fuels that could potentially help prolong the liquid fuels culture and mitigate the forthcoming effects of the shortage of transportation fuels. Biofuels have become more attractive recently because of their environmental benefits (Balat 2009; Demirbas 2009a; Hacisalihoglu 2009).

The share of biofuels in the automotive fuel market is expected to grow rapidly over the next decade. Biofuels could be peaceful energy carriers for all countries. They are renewable and available throughout the world. Policy makers will need to pay more attention to the implications for the transition to a biofuel economy. The concept of sustainable development embodies the idea of the interconnectedness and the balance between economic, social, and environmental concerns (Demirbas 2008).

The most common biofuels, such as ethanol from corn, wheat or sugar beet and biodiesel from oilseeds, are produced from classic food crops that require high-quality agricultural land for growth. However, bioethanol is a petrol additive/substitute that can be produced from plentiful, domestic, cellulosic biomass resources such as herbaceous and woody plants, agricultural and forestry residues, and a large portion of municipal and industrial solid waste streams. Production of bioethanol from biomass is one way to reduce both the consumption of crude oil and environmental pollution. There is also a growing interest in the use of vegetable oils for making biodiesel, which is less polluting than conventional petroleum diesel fuel.

Biofuels production costs can vary widely by feedstock, conversion process, scale of production and region. For biofuels, the cost of feedstock (crops) is a major component of overall costs. Total biofuel costs should also include a component representing the impact of biofuels production on related markets, such as food. In particular, the cost of producing oil-seed-derived biodiesel is dominated by the cost of the oil and by competition from high-value uses like cooking.

In 2004, 3.4 billion gallons of fuel ethanol were produced from over 10% of the corn crop. Ethanol demand is expected to more than double in the next ten years. For the supply to be available to meet this demand, new technologies must

Table 8.2 Economic and environmental impacts of biofuels

Economic impacts	Sustainability
	Fuel diversity
	Increased number of rural manufacturing jobs
	Increased income taxes
	Increased investments in plant and equipment
	Agricultural development
	International competitiveness
	Reducing the dependency on imported petroleum
Environmental impacts	Greenhouse gas reductions
	Reducing of air pollution
	Biodegradability
	Higher combustion efficiency
	Improved land and water use
	Carbon sequestration

be moved from the laboratories to commercial reality (Bothast 2005). The world ethanol production is about 60% from feedstock from sugar crops.

Biodiesel is a synthetic diesel-like fuel produced from vegetable oils, animal fats or waste cooking oil. It can be used directly as fuel, which requires some engine modifications, or blended with petroleum diesel and used in diesel engines with few or no modifications. At present, biodiesel accounts for less than 0.2% of the diesel consumed for transport. However, biodiesel has become more attractive recently because of its environmental benefits. The cost of biodiesel, however, is the main obstacle to commercialization of the product. With cooking oils used as raw material, the viability of a continuous transesterification process and recovery of high-quality glycerol as a biodiesel byproduct are primary options to be considered to lower the cost of biodiesel (Ma and Hanna 1999; UN 2006).

Some advantages of biofuels are the following: (a) biofuels are easily available from common biomass sources, (b) they are represent a carbon dioxide cycle in combustion, (c) they can be environmentally friendly, (d) there are many benefits to the economy and consumers in using biofuels, and (e) biofuels are biodegradable and contribute to sustainability (Puppan 2002). The benefits include greenhouse gas reductions including reduced carbon dioxide emissions, which will contribute to domestic and international targets, the diversification of the fuel sector, biodegradability, sustainability, and an additional market for agricultural products. Biofuels help to protect and create jobs. Table 8.2 shows the economic and environmental impacts of biofuels.

The land availability for producing biofuels in a well-established infrastructure makes an impact in the energy economy. The infrastructure contains a sustainable truck traffic system, where a well-established railroad system is available, trucks would only take the biomass to collection centers and then it would be railed to the biorefinery. Assuming that a high-yield sugar crop is planted all around the biorefinery, the longest trip a truck would need to make would be a 45 km round trip (Demirbas 2008).

8.1.1 Costs, Prices, and Economic Impacts of Biofuels

In previous economic studies of biodiesel production, the main economic factors such as capital cost, plant capacity, process technology, raw material cost and chemical costs were determined (Zhang et al. 2003). The major economic factor to consider for input costs of biodiesel production is the feedstock, which is about 75–80% of the total operating cost. Other important costs are labor, methanol and catalyst, which must be added to the feedstock (Balat and Balat 2008). Using an estimated process cost, exclusive of feedstock cost, of US $0.158/L for biodiesel production, and estimating a feedstock cost of US $0.539/L for refined soy oil, an overall cost of US $0.70/L for the production of soy-based biodiesel was estimated (Haas et al. 2006). Palm oil is the main option that is traded internationally, and with potential for import in the short term (Dene and Hole 2006). Costs for production from palm oil are estimated; the results are shown in Table 8.3 (Dene and Hole 2006).

The oil in vegetable seeds is converted into biodiesel through oil extraction, oil refining, and transesterification. The cost of biodiesel can be lowered by increasing feedstock yields, developing novel technologies, and increasing economic return on glycerol production by finding other uses for this byproduct, which, at the moment, due to oversupply is sold for little or no value. Alternatively, the use of co-solvents,

Table 8.3 Costs of biodiesel production

Plant size (million liters)	Capital costs	Feedstock	Methanol	Other	Glycerol credit	Distribution and blending	Total
Tallow fat (US $/L)							
6	0.33	0.40	0.05	0.11	0.12	0.08	0.85
23	0.15	0.40	0.05	0.10	0.12	0.08	0.66
46	0.11	0.40	0.05	0.09	0.12	0.09	0.61
69	0.08	0.40	0.05	0.08	0.12	0.15	0.64
137	0.06	0.40	0.05	0.06	0.12	0.15	0.60
Palm oil (US $/L)							
60	0.09	0.73	0.05	0.08	0.12	0.04	0.88
71	0.08	0.73	0.05	0.08	0.12	0.04	0.86
143	0.06	0.73	0.05	0.06	0.12	0.04	0.82

such as tetrahydrofuran, can make the alcohol-oil-ester-glycerol system into a single phase, thereby reducing the processing costs (Granda et al. 2007). However, these improvements still would not make biodiesel economically competitive at the current stage.

Biofuel production costs can vary widely by feedstock, conversion process, scale of production and region. On an energy basis, ethanol is currently more expensive to produce than gasoline in all regions considered. Only ethanol produced in Brazil comes close to competing with gasoline. Ethanol produced from corn in the US is considerably more expensive than from sugar cane in Brazil, and ethanol from grain and sugar beet in Europe is even more expensive. These differences reflect many factors, such as scale, process efficiency, feedstock costs, capital and labor costs, co-product accounting, and the nature of the estimates.

Average international prices for common biocrude, fat, crops and oils used as feedstock for biofuel production in 2007 are given in Table 8.4 (Demirbas 2009b). The cost of feedstock is a major economic factor in the viability of biodiesel production. Nevertheless, the price of waste cooking oil is 2.5–3.5 times cheaper than virgin vegetable oils, thus can significantly reduce the total manufacturing cost of biodiesel (Table 8.4). Biodiesel obtained from waste cooking vegetable oils has been considered a promising option. Waste cooking oil is available for a relatively cheap price for biodiesel production in comparison with fresh vegetable oil costs.

Economic advantages of a biofuel industry would include value added to the feedstock, an increased number of rural manufacturing jobs, an increase in income taxes, investments in plant and equipment, reduced greenhouse gas emissions,

Table 8.4 Average international prices for common biocrude, fat, crops and oils used as feedstock for biofuel production in 2007 (US $/ton)

Biocrude	167
Maize	179
Sugar	223
Wheat	215
Crude palm oil	543
Rapeseed oil	824
Soybean oil	771
Refined cottonseed oil	782
Crude corn oil	802
Crude peanut oil	891
Crude tea seed oil	514
Waste cooking oil	224
Yellow grease	412
Poultry fat	256

reduction in a country's reliance on crude oil imports, and supported agriculture by providing new labor and market opportunities for domestic crops. In recent years, the importance of non-food crops increased significantly. The opportunity to grow non-food crops under the compulsory set-aside scheme is an option to increase the biofuel production.

The socioeconomic impacts on the local economy arising from providing power through renewable resources instead of conventional generation technologies are very important. These impacts include direct and indirect differences in the jobs, income, and gross output. There are significant socioeconomic impacts associated with the investment in a new power plant, including increases in employment, output, and income in the local and regional economy. Increases in these categories occur as labor is directly employed in the construction and operation of a power plant, as local goods and services are purchased and utilized.

The potential for reduced costs of renewable liquid fuels and conservation of scarce fuel resources results in significant reductions in fuel usage. In addition to these economic benefits, development of renewable resources will have environmental, health, safety and other benefits.

Agriculture ethanol is at present more expensive than synthesis ethanol from ethylene. The simultaneous production of biomethanol (from sugar juice) in parallel to the production of bioethanol, appears to be economically attractive in locations where hydroelectricity is available at very low cost (~0.01 $kWh) (RFA 2007).

Currently there is no global market for ethanol. The crop types, agricultural practices, land and labor costs, plant sizes, processing technologies and government policies in different regions considerably vary ethanol production costs and prices by region. The cost of producing bioethanol in a dry mill plant currently totals US $1.65/gallon. Corn accounts for 66% of operating costs, while energy (electricity and natural gas) to fuel boilers and dry distillers dried grains represents nearly 20% of operating costs (Grassi 1999).

Ethanol from sugar cane, produced mainly in developing countries with warm climates, is generally much cheaper to produce than ethanol from grain or sugar beet in IEA countries. For this reason, in countries like Brazil and India, where sugar cane is produced in substantial volumes, sugar cane-based ethanol is becoming an increasingly cost-effective alternative to petroleum fuels. Ethanol derived from cellulosic feedstock using enzymatic hydrolysis requires much greater processing than from starch or sugar-based feedstock, but feedstock costs for grasses and trees are generally lower than for grain and sugar crops. If targeted reductions in conversion costs can be achieved, the total cost of producing cellulosic ethanol in OECD countries could fall below that of grain ethanol.

Estimates show that bioethanol in the EU becomes competitive when the oil price reaches US $70 a barrel, while in the US it becomes competitive at US $50–60 a barrel. For Brazil the threshold is much lower – between US $25 and US $30 a barrel. Other efficient sugar-producing countries such as Pakistan, Swaziland, and Zimbabwe have production costs similar to Brazil's (Urbanchuk 2007). Anhydrous ethanol, blendable with gasoline, is still somewhat more expensive. Prices in India have declined and are approaching the price of gasoline.

Table 8.5 The top ten bioethanol producers (billion gallons)

Country	2004	2005	2006
United States	3.54	4.26	4.85
Brazil	3.99	4.23	4.49
China	0.96	1.00	1.02
India	0.46	0.45	0.50
France	0.22	0.24	0.25
Germany	0.07	0.11	0.20
Russia	0.20	0.20	0.17
Canada	0.06	0.06	0.15
South Africa	0.11	0.10	0.10
Thailand	0.07	0.08	0.09

Generally, the larger US conversion plants produce biofuels, particularly ethanol, at lower cost than plants in Europe. Production costs for ethanol are much lower in countries with a warm climate, with Brazil probably the lowest-cost producer in the world. Production costs in Brazil, using sugar cane as the feedstock, have recently been recorded at less than half the costs in Europe. Production of sugar cane ethanol in developing countries could provide a low-cost source for substantial displacement of oil worldwide over the next 20 years. Table 8.5 shows the top ten bioethanol producers (RFA 2007).

Biorefineries of the future will produce a wide array of products derived from biomass that cannot even be envisioned at this time. These products will include specialty chemicals for the chemical manufacturing industry, polymers, and fuels. However, current markets are not yet ready to accept the majority of the products that will become available through biorefineries. In the near term, biorefineries must rely on products that already have a market in today's economy. Production of fuel-grade ethanol is the key to initial economic survival for biorefineries. Fuel-grade ethanol has a current market for use as an oxygenated additive to motor gasoline and potential for use as an octane booster in near-term applications.

The largest ethanol cost component is the plant feedstock. Operating costs, such as feedstock cost, co-product credit, chemicals, labor, maintenance, insurance and taxes, represent about one-third of total cost per liter, of which the energy needed to run the conversion facility is an important (and in some cases quite variable) component. Capital cost recovery represents about one-sixth of total cost per liter. It has been showed that plant size has a major effect on cost (Dufey 2006). The plant size can reduce operating costs by 15 to 20%, saving another $0.02 to $0.03 per liter. Thus, a large plant with production costs of $0.29 per liter may be saving $0.05 to $0.06 per liter over a smaller plant (Whims 2002).

Biodiesel from animal fat is currently the cheapest option ($0.4 to $0.5 per liter) while traditional transesterification of vegetable oil is at present around $0.6 to $0.8 per liter (IEA 2007). Rough projections of the cost of biodiesel from vegetable oil and waste grease are, respectively, \$0.54 to \$0.62 per liter and \$0.34 to \$0.42 per liter. With pretax diesel priced at \$0.18 per liter in the US and \$0.20 to 0.24 per liter in some European countries, biodiesel is thus currently not economically feasible, and more research and technological development will be needed (Bender 1999).

The calculation of biofuel prices should be designed to maintain into the future the equilibrium between supply and demand, taking into account the costs of planned investments. They should also take into account the rest of the economy and the environment. Two very important characteristics of energy prices are equity and affordability. Biofuel prices must reflect the cost, imposed by the specific consumer category on the economy. Since energy prices based on apparent long-run marginal costs may not be sufficient to finance the development of the energy sector, the prices should be adjusted, so that the energy sector can be financed without subsidies to enhance its autonomy. In competitive markets this form of adjustment may not be possible.

The marginal biofuel is the most expensive fuel that would be used where a target was met at least cost. Fuels would be used in turn in order of price from lowest to highest. Our concern is with the most expensive fuel that would be used to meet the target, as that is the one that would be displaced by a lower-cost fuel. Manufacture from wood is untested commercially at this stage although at least one plant is expected to come into production in 2006.

There are three cost factors that affect the biofuel costs:

1. Operating costs
2. Distribution and blending costs
3. Capital costs

The operating costs are defined here as all costs that are not capital or feedstock costs. The literature contains large variations in the estimation of operating costs. Where there is a range of costs reported in different sources, a midpoint has been used. There are two components to distribution costs: the costs of delivering the feedstock to the ethanol production plant, and the costs of distributing the ethanol to the blending facility. The feedstock cost consists of two parts: the conversion yield (how much ethanol is produced from one ton of feedstock), and the price of that feedstock. Once again, there are large variations in the literature in the estimation of conversion yield. In these cases, a midpoint has been used.

References

Balat, M., Balat, H. 2008. A critical review of bio-diesel as a vehicular fuel. Energy Convers Manage 49:2727–2741.

Balat, M. 2009. New biofuel production technologies. Energy Edu Sci Technol Part A 22:147–161.

Bender, M. 1999. Economic feasibility review for community-scale farmer cooperatives for biodiesel. Biores Technol 70:81–87.

Bothast, R.J. 2005. New technologies in biofuel production. Agricultural Outlook Forum (AOF), 24 Feb 2005. http://www.usda.gov/oce/forum/speeches/Bothast.pdf. Accessed 2009.

Demirbas, A. 2008. Economic and environmental impacts of the liquid biofuels. Energy Edu Sci Technol 22:37–58.

Demirbas, C. 2009a. The global climate challenge: recent trends in CO_2 emissions from fuel combustion. Energy Edu Sci Technol Part A 22:179–193.

Demirbas, A.H. 2009b. Inexpensive oil and fats feedstocks for production of biodiesel. Energy Edu Sci Technol Part A 23:1–13.

Dene, T., Hole, J. 2006. Enabling biofuels: biofuel economics. Final Report. Ministry of Transport, Auckland, New Zealand.

Dufey, A. 2006. Biofuels production, trade and sustainable development: emerging issues. Environmental Economics Programme, Sustainable Markets Discussion Paper No. 2. International

Institute for Environment and Development (IIED), London, September, 2006.

Granda, C.B., Zhu, L., Holtzapple, M.T. 2007. Sustainable liquid biofuels and their environmental impact. Environ Progress 26:233–250.

Grassi, G. 1999. Modern bioenergy in the European Union. Renew Energy 16:985–990.

Haas, M.J., McAloon, A.J., Yee, W.J., Foglia, T.A. 2006. A process model to estimate biodiesel production costs. Biores Technol 97:671-678.

Hacisalihoglu, S. 2009. Ethanol-gasoline and ethanol-diesel fuel blends. Energy Edu Sci Technol Part A 22:133–146.

IEA (International Energy Agency). 2007. Biodiesel statistics. IEA Energy Technology Essentials, OECD/IEA, Paris, January, 2007.

Ma, F., Hanna, M.A. 1999. Biodiesel production: a review. Biores Technol 70:1–15.

Puppan, D. 2002. Environmental evaluation of biofuels. Periodica Polytechnica Ser Soc Man Sci 10:95–116.

UN (United Nations). 2006. The emerging biofuels market: regulatory, trade and development implications. United Nations Conference on Trade and Development, New York and Geneva.

Zhang, Y., Dub, M.A., McLean, D.D., Kates, M. 2003. Biodiesel production from waste cooking oil: 2. Economic assessment and sensitivity analysis. Biores Technol 90:229–240.

RFA (Renewable Fuels Association). 2007. Ethanol Industry Statistics. RFA, Washington, DC.

Urbanchuk, J.M. 2007. Economic impacts on the farm community of cooperative ownership of ethanol production. United States Department of Agriculture, Agricultural Outlook Forum, No. 8083 2007.

Whims, J. 2002. Corn based ethanol costs and margins. Attachment 1, AGMRC, Kansas State University, 2002. http://www.agmrc.org/corn/info/ksueth1.pdf. Accessed 2009.

Chapter 9
Political Impacts of Biorefinery

9.1 Introduction

Current energy policies address environmental issues including environmentally friendly technologies to increase energy supplies and encourage cleaner, more efficient energy use, and address air pollution, the greenhouse effect, global warming, and climate change (Demirbas 2006). The biofuel policy aims to promote the use of transport fuels made from biomass, as well as other renewable fuels. Biofuels potentially provide new economic opportunities for people in rural areas in oil importer and developing countries. The central policy of biofuel concerns job creation, greater efficiency in the general business environment, and protection of the environment. Projections are important tools for long-term planning and policy settings. Renewable energy sources that use indigenous resources have the potential to provide energy services with zero or almost zero emissions of both air pollutants and greenhouse gases. Biofuels are expected to reduce dependence on imported petroleum with associated political and economic vulnerability, reduce greenhouse gas emissions and other pollutants, and revitalize the economy by increasing demand and prices for agricultural products.

In general, energy policy includes issues of energy production, distribution and consumption. The attributes of energy policy may include international treaties, legislation on commercial energy activities (trading, transport, storage, etc.), incentives for investment, guidelines for energy production, conversion and use (efficiency and emission standards), taxation and other public policy techniques, energy-related research and development, energy economy, general international trade agreements and marketing, energy diversity, and risk factors of a possible energy crisis.

Today's world is facing three critical problems: (1) high fuel prices, (2) climatic changes, and (3) air pollution. Currently there are several important problems to be resolved worldwide: (1) high need for energy, (2) high depletion of non-renewable energy recourses, and (3) high local and global environmental pollution (Demirbas 2009a). The use of biofuels has many advantages including energy security reasons, environmental concerns, foreign exchange savings, and socioeconomic issues related to the rural sector. The biggest difference between biofuels and petroleum feedstocks is oxygen content. Biofuels have oxygen levels from 10 to 45%, while petroleum has essentially none, making the chemical properties of biofuels very different from petroleum. Oxygenates are just preused hydrocarbons having structure that provides a reasonable anti-knock value (Demirbas 2009b).

There is considerable uncertainty in the economic analysis, associated with the costs of feedstocks and oil prices, and also because of the rapid development of the technologies. Any policy to support and encourage the supply of biofuel should provide incentives for both biodiesel and ethanol supplies. Policy options include incentive payments or tax breaks. Due to rising prices for fossil fuels (especially oil, but also natural gas and to a lesser extent coal) the competitiveness of biomass use has improved considerably over time. Biomass and bioenergy are now a key option in energy policies. Security of supply, an alternative for mineral oil and reduced carbon emissions are key reasons for the use of biomass for energy. Targets and expectations for bioenergy in many national policies are ambitious, reaching 20–30% of total energy demand in various countries. Similarly, long-term energy scenarios also contain challenging targets.

9.2 Political Impacts

Energy is an essential input driving economic development. Therefore, in developed economies energy policies constitute an important component of overall regulatory frameworks, shaping and improving overall competitiveness and market integration of the private business sector. Overall competitiveness includes liberalization of the electricity and gas markets, as well as by the separation of energy production, transportation, and distribution activities.

Policy drivers for renewable liquid biofuels have attracted particularly high levels of assistance in some countries given their promise of benefits in several areas of interest to governments, including agricultural production, greenhouse gas emissions, energy security, trade balances, rural development and economic opportunities for developing countries. Total biofuel support estimates for US and EU in 2006 is given in Table 9.1 (Demirbas 2008).

Table 9.1 Total biofuel support estimates for US and EU in 2006 (billions of US $)

	Biodiesel	Ethanol	Total biofuels
United States	0.60	5.96	6.70
Europe Union (EU)	3.11	1.61	4.82
Global total	3.65	7.85	11.79

Table 9.2 Biorefinery policy and impacts depending on the priorities

	Policy priority	
	Energy security and agricultural policy	Efficiency and costs to consumer
Policies	Tariff and quota, subsidies, export tax on raw material, standards	Technology, international agreements, risk investment and R&D support
Positive impacts	Domestic production, security, rural employment and income, foreign policy and military costs	Efficient resource allocation and cost efficiency, less administrative costs, more predictable operation environment
Negative impacts	Trade wars, complex and manifold regulations, regional differences, uncertainty, resource misallocations	Loss of domestic production and rural deployment, imported energy dependency

A biorefinery is a facility that integrates biomass conversion processes and equipment to co-produce food, feed, materials, chemicals, power and/or heat from biomass. By producing multiple products, a biorefinery can take advantage of the differences in biomass components and intermediates and maximizes the value derived from the biomass feedstock. Because of concerns about the current first-generation of liquid biofuels, there is major interest in moving to alternative systems of biofuel production, such as second-generation liquid biofuels based on lignocellulosic biomass, and applications of such biotechnologies will be important if they are to become widely available in the future.

Policies designed to support the promotion of second-generation biofuels must be carefully developed if they are to avoid unwanted consequences and potentially delay commercialization. Deployment policies generally fall into two categories: blending targets, which can be mandatory or voluntary, and tax credits. Mandatory targets give certainty over outcomes, but not over the potential costs, while it is the inverse for tax credits. What pathways individual countries choose will depend critically on their policy goals and the risks they perceive. Deployment policies are essential if rapid scale-up of the industry is required to reduce costs through learning-by-doing. Otherwise, deployment and cost reductions are likely to be slow since initial commercial deployment focuses on niche opportunities where costs and risks are low. Table 9.2 shows the biorefinery policy and impacts depending on the priorities.

Many governments are actively encouraging liquid biofuel production in their countries. A number of policy mechanisms are being used by governments for this purpose, including feed-in tariffs (where a regulatory minimum guaranteed price is paid for renewable power fed into the electricity grid), taxes, guaranteed markets, compulsory grid connections for renewable energy producers and other direct support for bioenergy production, such as grants, loans and subsidies (GBEP 2007). Government policies are therefore playing a key role in influencing investment in bioenergy. Following GBEP (2007), there are four main factors driving the current interest in liquid biofuels:

1. High energy prices
2. Energy security
3. Climate change
4. Rural development

Current European Union (EU) policies on alternative motor fuels focus on the promotion of biofuels. The definition of the marginal producer depends on the policy stance on biofuels. Biofuel pricing policy should not be employed as an anti-inflationary instrument, instead it should be applied in such a way, that it does not create cross subsidies between classes of consumers. In a proposed biofuels directive the introduction of a mandatory share scheme for biofuels, including from 2009 minimum blending shares. Table 9.3 shows the shares of alternative fuels compared to the total automotive fuel consumption in the EU under the optimistic development scenario of the European Commission (Demirbas 2008). The EU has set the goal in February 2006 of fulfilling 5.75% of transportation fuel needs from biofuels by 2010 in all member states. In the commission's view mandating the use of biofuels will (a) improve energy supply security, (b) reduce greenhouse gas (GHG) emissions, and (c) boost rural incomes and employment (Jansen 2003; Hansen et al. 2005). The **European Union** accounted for nearly 89% of all biodiesel production worldwide in 2005. By 2010, the US is expected to become the world's largest single biodiesel market, accounting for roughly 18% of world biodiesel consumption, followed by Germany.

The general EU policy objectives considered most relevant to the design of energy policy are (Jansen 2003):

1. Competitiveness of the EU economy
2. Security of energy supply
3. Environmental protection

Policy interventions, especially in the form of subsidies and mandated blending of biofuels with fossil fuels, are driving the rush to liquid biofuels. For example, the EU decided in March 2007 to set mandatory targets of a 20% share of renewable energies in overall EU energy consumption by the year 2020, and a mandatory 10% minimum target for the share of biofuels in overall EU transport petrol and diesel consumption by 2020 (FAO 2008).

Table 9.3 Shares of alternative fuels in total automotive fuel consumption in the EU under the optimistic development scenario of the European Commission

Year	Biofuel	Natural gas	Hydrogen	Total
2010	6	2	–	8
2015	7	5	2	14
2020	8	10	5	23

Producing and using biofuels for transportation offers alternatives to fossil fuels that can help provide solutions to many environmental problems. Using biofuels in motor vehicles helps reduce green house gases (GHG) emissions. Biodiesel and ethanol provide significant reductions in GHG emissions compared to gasoline and diesel fuel. Due to the low or zero content of pollutants such as sulfur in biofuels, the pollutant emission of biofuels is much lower than the emission of conventional fuels. Numerous low-emission scenarios have demonstrated that the Kyoto Protocol cannot be achieved without establishing a large role for biofuel in the global energy economy by 2050. Low-emission scenarios imply 50–70 EJ of biofuel raw material in 2050. Well-designed biofuel projects would have very significant sustainable development benefits for rural areas, including creation of rural employment, rural electricity supply, soil conservation and environmental benefits (Demirbas 2006).

The main biofuel opportunities where suitable land is available are in developing countries. The issue of energy security has been accorded top-most priority. Every effort needs to be made to enhance the indigenous content of energy in a time-bound and planned manner. The additional benefit of biofuel development is creation of new employment opportunities in manufacturing, construction, plant operation and servicing, and fuel supply. Rural jobs are created in fuel harvesting, transport, and maintenance of processing areas (Demirbas 2006).

Biodiesel is a synthetic diesel-like fuel produced from vegetable oils, animal fats or waste cooking oil. It can be used directly as fuel, which requires some engine modifications, or blended with petroleum diesel and used in diesel engines with few or no modifications. At present, biodiesel accounts for less than 0.2% of the diesel consumed for transport. The cost of biodiesel, however, is the main obstacle to commercialization of the product. With cooking oils used as raw material, the viability of a continuous transesterification process and recovery of high-quality glycerol as a biodiesel byproduct are primary options to be considered to lower the cost of biodiesel. The possible impact of biodiesel on fuel economy is positive as given in Table 9.4 (EPA 2002).

Many farmers who raise oilseeds use a biodiesel blend in tractors and equipment as a matter of policy, to foster production of biodiesel and raise public awareness. It is sometimes easier to find biodiesel in rural areas than in cities. Additional factors must be taken into account, such as the fuel equivalent of the energy required for processing, the yield of fuel from raw oil, the return on cultivating food, and the relative cost of biodiesel versus petrodiesel. Biodiesel policy should reflect that theme by aggressively eliminating the government-imposed barriers to its success. Biodiesel policy should be based firmly in the philosophy of free-

Table 9.4 Fuel economy impacts of biodiesel use

Percentage of biodiesel in diesel fuel	Percentage reduction in miles per gallon
20	0.9–2.1
100	4.6–10.6

dom. Given the history of petroleum politics, it is imperative that today's policy decisions ensure a free market for biodiesel. Producers of all sizes must be free to compete in this industry, without farm subsidies, regulations and other interventions skewing the playing field. The production and utilization of biodiesel is facilitated, firstly through the agricultural policy of subsidizing the cultivation of non-food crops, and secondly, biodiesel is exempt from the oil tax (USDE 2006; Wardle 2003; USDA 2003).

References

Demirbas, A. 2006. Global biofuel strategies. Energy Edu Sci Technol 17:27–63.
Demirbas, A. 2008. Biodiesel: a realistic fuel alternative for diesel engines. Springer, London.
Demirbas, A. 2009a. Energy concept and energy education. Energy Edu Sci Technol Part B 1:85–101.
Demirbas, A. 2009b. Future Energy Sources: Part I. Future Energy Source 1:1–95.
EPA (US Environmental Protection Agency). 2002. A comprehensive analysis of biodiesel impacts on exhaust emissions. Draft Technical Report, EPA420-P-02-001, October 2002.
FAO. 2008. The State of Food and Agriculture 2008: Biofuels: prospects, risks and opportunities. Food and Agricultural Organization of the UN, Rome.
GBEP (Global Bioenergy Partnership). 2007. A review of the current state of bioenergy development in G8 +5 countries. GBEP, New York.
Hansen, A.C., Zhang, Q., Lyne, P.W.L. 2005. Ethanol-diesel fuel blends: a review. Biores Technol 96:277–285.
Jansen, J.C. 2003. Policy support for renewable energy in the European Union. Energy Research Centre of the Netherlands. http://www.ecn.nl/docs/library/report/2003/C03113. Accessed 2009.
USDA (United States Department of Agriculture). 2003. Production estimates and crop assessment Division Foreign Agricultural Service. EU: Biodiesel Industry Expanding Use of Oilseeds. USDA, Washington, DC.
USDE (US Department of Energy). 2006. Biodiesel handling and use guidelines. DOE/GO-102006 2358, 3rd edn. USDE, Oak Ridge, TN.
Wardle, D.A. 2003. Global sale of green air travel supported using biodiesel. Renew Sust Energy Rev 7:1–64.

Chapter 10
Environmental Impacts of Biorefineries

10.1 Introduction

Biomass is a sustainable alternative to fossil energy carriers, which are used to produce fuels, electricity, chemicals, and other goods. Biorefineries are supposed to contribute to a more sustainable resource supply and to a reduction in greenhouse gas emissions. However, bio-based products and fuels may also be associated with environmental disadvantages due to, e. g., land use or eutrophication of water (eutrophic waters extremely rich in nutrients). The biorefinery has the greatest environmental impact on three categories: fossil fuel use, respiratory effects, and carcinogenic substances. The environmental impacts predominantly result from the provision of hydrochloric acid and to a smaller extent also from the provision of process heat. The use of biofuels to reduce greenhouse gas (GHG) emissions and fossil energy depletion is being increasingly contested due to overall sustainability concerns.

A number of studies with comparisons of (1) gasoline, ethanol and gasoline/ethanol blends, (2) diesel, ethanol and diesel/ethanol blends, and (3) diesel, natural gas and diesel/biodiesel blends of bus emissions have been published previously (ARS 2002; Kim and Choi 2008; Yu and Tao 2009). The basis for these comparisons, the choice of vehicles and even the outcome vary significantly. The studies have been conducted to develop a methodology for the generation of driving and duty cycles for refuse vehicles matching the statistical metrics and distributions of the generated cycles to the collected database (Rakopoulos et al. 2008; Lapuerta et al. 2008). These cycles can be utilized in the development and computer simulation of future refuse vehicle designs, specifically liquid biofuel-fueled vehicles. Other comparisons include a chassis-based testing program to document emission reduction (NO_x, SO_x, HC, CO, and PM) performance from a fleet of vehicles accumulating mileage or applying standard driving cycles (Graham et al. 2008; Corro and Ayala 2008). Methodology is also based on land use changes over time. A certain amount of land

Table 10.1 General overview of methodology and data requirement for biofuels

INPUTS			OUTPUTS		
Design	Cost factors	Data analysis	Political impact	Economic impact	Environmental impact
Literature review	Capital cost	Cost data	Benefits	Combustion efficiency	Emission reducing
	Feedstock price				
	Chemicals				
Raw materials	Investment support	Scenarios			
	Process				
	Land use	Future view			
Comparisons	Subsidy and credit				

is needed to meet the required production for biofuel raw materials. A general overview of methodology and data requirement for biofuels is shown in Table 10.1.

Biofuels have become more attractive recently due to their environmental benefits. Ethanol addition to gasoline has increased the engine torque, power, and fuel consumption and reduced carbon monoxide (CO) and hydrocarbon (HC) emissions. In general, biodiesel increases NO_x emissions when used as fuel in a diesel engine. The emissions of biodiesel (B20 and B100) for same model compression ignition (diesel) vehicles have increased from 1.86 to 2.23, respectively. The fact that NO_x emissions increase with increasing biodiesel concentration could be a detriment in areas that are out of attainment for ozone. Oxygenated diesel fuel blends have a potential to reduce the emission of particulate matter (PM) and to be an alternative to diesel fuel.

10.2 Environmental Impacts

The goal in biorefining is to isolate all the added value from the biomass feedstock, resulting in little or no waste. This will not only improve the economics so that such processes can compete with the petrochemical industry, but will also lower the overall environmental impact.

The combined production methods will significantly simplify the overall production process and thereby substantially reduce the product costs and the environmental impacts.

A number of studies with comparisons of diesel, natural gas and diesel/biodiesel blends have been published previously (ARS 2002; Janulis 2004; Tzirakis et al. 2007; Krahl et al. 2009; Soltic et al. 2009; Coronado et al. 2009). Biodiesel has a good energy return because of the simplicity of its manufacturing process. It also has significant benefits in emissions and could also play an important role in the energy economy if higher crop productivities are attained (Granda et al. 2007).

Table 10.2 Emissions of biodiesel for same-model diesel vehicles

Vehicle or engine	Fuel	Emissions (g/km)				
		NO_x	CO	CH	PM	SO_x
Peugeot Partner	B100	2.05	9.37	0.54	2.68	0
Peugeot Partner	B20	1.86	17.73	1.32	4.71	0.004
Renault Kangoo	B100	2.23	9.22	0.49	3.06	0
Renault Kangoo	B20	1.92	17.36	1.26	5.63	0.003
Dacia Pickup	B100	2.15	9.42	0.56	2.59	0
Dacia Pickup	B20	1.91	18.29	1.35	4.63	0.005

Table 10.3 Major environmental impacts of biorefining fuels

Greenhouse gas reductions
Reduction of air pollution
Biodegradability
Higher combustion efficiency
Improved land and water use
Carbon sequestration
Lower sulfur content
Lower aromatic content
Less toxicity

Table 10.2 shows the emissions of biodiesel (B20 and B100) for same-model compression ignition (diesel) vehicles (Demirbas 2009). Emissions of NO_x increase with increasing biodiesel amount in the blends. The properties of biodiesel and diesel fuels, in general, show many similarities, and therefore, biodiesel is rated as a realistic fuel as an alternative to diesel. There are several ways to control NO_x in a biodiesel engine: run the engine very lean, which lowers the temperature, or very rich, which reduces the oxygen supplies, or decrease the burn time or lower the engine. Emissions of NO_x also increase with the combustion temperature, the length of the high-temperature combustion period, and the availability of biodiesel, up to a point. Major environmental impacts of biorefining fuels are shown in Table 10.3.

Alcohols have been used as a fuel for engines since nineteenth century. Among the various alcohols, ethanol is known as the best-suited renewable, bio-based and ecofriendly fuel for spark-ignition (SI) engines. The most attractive properties of ethanol as an SI engine fuel are that it can be produced from renewable energy sources such as sugar, cane, cassava, many types of waste biomass materials, corn and barley. In addition, ethanol has higher evaporation heat, octane number and flammability temperature therefore it has a positive influence on engine performance and reduces exhaust emissions. The results of the engine test showed that ethanol addition to unleaded gasoline increase the engine torque, power and fuel consumption and reduces carbon monoxide and hydrocarbon emissions.

The biodiesel impacts on exhaust emissions varied depending on the type of biodiesel and on the type of conventional diesel. Blends of up to 20% biodiesel mixed with petroleum diesel fuels can be used in nearly all diesel equipment and are compatible with most storage and distribution equipment. Using biodiesel in a conventional diesel engine substantially reduces emissions of unburned hydrocarbons, carbon monoxide, sulfates, polycyclic aromatic hydrocarbons, nitrated polycyclic aromatic hydrocarbons, and particulate matter. These reductions increase as the amount of biodiesel blended into diesel fuel increases. In general, biodiesel increases NO_x emissions when used as fuel in diesel engine. The pollutant emissions of ethanol–gasoline blends of 0%, 5%, 10%, 15% and 20% were experimentally analyzed in a four-stroke engine. The concentration of CO and hydrocarbon emissions in the exhaust pipe were measured and found to decrease when ethanol blends were introduced. This was due to the high oxygen percentage in the ethanol. In contrast, the concentration of CO_2 and NO_x was found to increase when ethanol was introduced.

Oxygenated diesel fuel blends have a potential to reduce the emission of particulate matter (PM) and to be an alternative to diesel fuel. Results obtained showed that the addition of bioethanol to the diesel fuel may be necessary to decrease diesel PM generation during combustion (Yu and Tao 2009; Corro and Ayala 2008). The total number and total mass of the PM of the ethanol–diesel blend fuels were decreased by about 11.7–26.9% (Kim and Choi 2008).

An experimental investigation was conducted to evaluate the effects of using blends of ethanol with conventional diesel fuel, with 5% and 10% (by volume) ethanol, on the performance and exhaust emissions of a fully instrumented, six-cylinder, turbocharged and after-cooled, heavy duty, direct injection (DI), Mercedes-Benz engine. Fuel consumption, exhaust smokiness and exhaust regulated gas emissions such as nitrogen oxides, carbon monoxide and total unburned hydrocarbons were measured. The differences in the measured performance and exhaust emissions of the two ethanol–diesel fuel blends from the baseline operation of the engine, i. e., when working with neat diesel fuel, were determined and compared (Rakopoulos et al. 2008). Diesel emissions were measured from an automotive engine using anhydrous bioethanol blended with conventional diesel, with 10% ethanol in volume and no additives. The resulting emissions were compared with those from pure diesel (Lapuerta et al. 2008).

The results of the statistical analysis suggest that the use of E10 results in statistically significant decreases in CO emissions (–16%), statistically significant increases in emissions of acetaldehyde (108%), 1,3-butadiene (16%), and benzene (15%), and no statistically significant changes in NO_x, CO_2, CH_4, N_2O, or formaldehyde emissions. The statistical analysis suggests that the use of E85 results in statistically significant decreases in emissions of NO_x (–45%), 1,3-butadiene (–77%), and benzene (–76%), statistically significant increases in emissions of formaldehyde (73%) and acetaldehyde (2540%), and no statistically significant change in CO and CO_2 emissions (Graham et al. 2008).

Combustion is the chemical reaction of a particular substance with oxygen, where there is a combination of inflammable matter with oxygen accompanied by release of heat. The quantity of heat evolved when one mole of a hydrocarbon is burned to carbon dioxide and water is called the heat of combustion. This type of combustion to carbon dioxide and water is characteristic of organic compounds and used under special conditions to determine the carbon and hydrogen content of a substance (Demirbas 2006a). During combustion, the combustible part of the fuel is divided into the volatile part and the solid residue. During heating it evaporates together with a part of carbon in the form of hydrocarbon combustible gases and carbon monoxide release by thermal degradation of the fuel. Carbon monoxide is mainly formed in the following reactions: (a) from reduction of CO_2 with unreacted C ($CO_2+C \rightarrow 2CO$), and (b) from degradation of carbonyl fragments (–CO) in the fuel molecules at 600–750 K.

The combustion process begins by heating the fuel above its ignition temperature in the presence of oxygen or air. Under the influence of heat, the chemical bonds of the fuel are cleaved. If complete combustion occurs, the combustible elements (C, H and S) react with the oxygen content of the air to form CO_2, H_2O and mainly SO_2.

If not enough oxygen is present or the fuel and air mixture is insufficient then the burning gases are partially cooled below the ignition temperature and the combustion process stays incomplete. In terms of environmental effects, the flue gases that escape to the atmosphere then still contain combustible components, mainly carbon monoxide (CO), unburned carbon (C) and various hydrocarbons (C_xH_y).

The standard measure of the energy content of a fuel is its heating value (HV), sometimes called the calorific value or heat of combustion. In fact, there are multiple values for the HV, depending on whether it measures the enthalpy of combustion (ΔH) or the internal energy of combustion (ΔU), and whether a fuel containing hydrogen product water is accounted for in the vapor phase or the condensed (liquid) phase. With water in the vapor phase, the lower heating value (LHV) at constant pressure measures the enthalpy will change due to combustion (Jenkins et al. 1998). The heating value is obtained by the complete combustion of a unit quantity of solid fuel in an oxygen-bomb calorimeter under carefully defined conditions. The gross heat of combustion or higher heating value (GHC or HHV) is obtained by the oxygen-bomb calorimeter method as the latent heat of moisture in the combustion products is recovered.

Biofuels are oxygenated compounds, e. g., ethanol, methanol, and biodiesels, and they provide more efficient combustion, and cleaner emissions. At a stoichiometric air/fuel ratio of 9:1 in comparison with gasoline's 14.7:1, it is obvious that more ethanol is required to produce the chemically correct products of CO_2 and water.

Ethanol has a higher octane number (108), broader flammability limits, higher flame speeds and higher heats of vaporization than gasoline. These properties allow for a higher compression ratio, shorter burn time and leaner burn engine, which lead to theoretical efficiency advantages over gasoline in an internal combustion engine. The octane number of ethanol allows it to sustain significantly higher internal pressures than gasoline, before being subjected to predetonation. Disadvantages of ethanol include its lower energy density than gasoline, its corrosiveness, low flame luminosity, lower vapor pressure, miscibility with water, and its toxicity to ecosystems.

Methanol is another possible fuel, since its higher octane number of methyl (114) alcohol would contribute to an increase in the engine compression ratio. This would increase the efficiency of converting the potential combustion energy to power. Finally, alcohols burn more completely, thus increasing combustion efficiency.

Biofuels such as bioethanol, biomethanol, biohydrogen and biodiesel generally have lower emissions than fossil-based engine fuels. Many studies on the performances and emissions of compression ignition engines, fueled with pure biodiesel and blends with diesel oil, have been performed and are reported in the literature.

Another potential fuel source is vegetable oils, which have become more attractive recently because of their environmental benefits and the fact that they are made from renewable resources. Dorado et al. (2003) describe experiments on the exhaust emissions of biodiesel from olive oil methyl ester as an alternative to diesel fuel in a diesel direct injection Perkins engine.

The methyl ester of vegetable oil was evaluated as a fuel in a compression ignition engine by researchers (Dunn 2001). They concluded that the performance of the esters of vegetable oil did not differ greatly from that of diesel fuel. The brake power was nearly the same as with diesel fuel, while the specific fuel consumption was higher than that of diesel fuel. Based on crankcase oil analysis, engine wear rates were low but some oil dilution did occur. Carbon deposits inside the engine were normal, with the exception of intake valve deposits.

The results showed the transesterification treatment decreased the injector coking to a level significantly lower than that observed with diesel fuel (Shay 1993). Although most researchers agree that vegetable oil ester fuels are suitable for use in a compression ignition engine, a few contrary results have also been obtained. The results of these studies point out that most vegetable oil esters are suitable as diesel substitutes but that more long-term studies are necessary for commercial utilization to become practical.

The use of biodiesel to reduce N_2O is attractive for several reasons. First, biodiesel contains little nitrogen, as compared with diesel fuel, which is also used as a reburning fuel. The N_2O reduction was strongly dependent on initial N_2O concentration and only slightly dependent upon temperature, where increased temperature increased N_2O reduction. This results in lower N_2O production from fuel nitrogen species for biodiesel. In addition, biodiesel contains virtually trace amount of sulfur, so SO_2 emissions are reduced in direct proportion to the diesel fuel replacement. Biodiesel is a regenerable fuel; when a diesel fuel is replaced by a biodiesel, there is a net reduction in CO_2 emissions. As an energy source used in a diesel engine, biodiesel reduces the consumption of diesel fuels and thereby reduces greenhouse gas emissions. Additional effects are a reduction of the ash volume and the SO_x and NO_x emissions of a biodiesels-fueled compression ignition engine. Neat biodiesel and biodiesel blends reduce particulate matter, hydrocarbons and carbon monoxide emissions and increase nitrogen oxide (NO_x) emissions compared with diesel fuel used in an unmodified diesel engine (Grassi 1999). The emission impacts of 20% by volume soybean-based biodiesel added to an average base petrodiesel is given in Table 10.4 (Balat and Balat 2008).

Table 10.4 Emission impacts of 20% by volume soybean-based biodiesel added to an average base petrodiesel

	Percent change in emissions
NO_x (nitrogen oxides)	+2.0
Particular matter	−10.1
Hydrocarbons	−21.1
Carbon monoxide	−11.0

The total net emission of carbon dioxide (CO_2) is considerably less than that of diesel oil and the amount of energy required for the production of biodiesel is less than that obtained with the final product. In addition, the emission of pollutants is somewhat less. CO_2, one of the primary greenhouse gases, is a transboundary gas, which means that, after it is emitted by a source, it is quickly dispersed in our atmosphere by natural processes. The use of biodiesel reduces CO_2 emissions. Table 10.5 shows the average biodiesel emissions compared to conventional diesel, according to EPA (2002). Table 10.6 shows the average changes in mass emissions from diesel engines using the biodiesel mixtures relative to the standard diesel fuel.

Results indicate that the transformities of biofuels are greater than those of fossil fuels, thus showing that a larger amount of resources is required to obtain an envi-

Table 10.5 Average biodiesel emissions (%) compared to conventional diesel

Emission type	Pure biodiesel	20% Biodiesel +80% Petrodiesel
	B100	B20
Total unburned hydrocarbons (HC)	−67	−20
Carbon monoxide	−48	−12
Particulate matter	−47	−12
NO_x	+10	+2
Sulfates	−100	−20
Polycyclic aromatic hydrocarbons[a]	−80	−13
Ozone potential of speciated hydrocarbons	−50	−10

[a]Average reduction across all compounds measured

Table 10.6 Average changes in mass emissions from diesel engines using the biodiesel mixtures relative to the standard diesel fuel (%)

Mixture	CO	NO_x	SO_2	Particular matter	Volatile organic compounds
B20	−13.1	+2.4	−20	−8.9	−17.9
B100	−42.7	+13.2	−100	−55.3	−63.2

ronmentally friendly product. This can be explained by the fact that natural processes are more efficient than industrial ones. On the other hand, the time involved in the formation of the fossil fuels is considerably different from that required for the production of the biomass (Carraretto et al. 2004).

Different scenarios for the use of agricultural residues as fuel for heat or power generation are analyzed. Reductions in net CO_2 emissions are estimated at 77–104 g/MJ of diesel displaced by biodiesel. The predicted reductions in CO_2 emissions are much greater than values reported in recent studies on biodiesel derived from other vegetable oils, due both to the large amount of potential fuel in the residual biomass and to the low-energy inputs in traditional coconut farming techniques. Unburned hydrocarbon emissions from biodiesel fuel combustion decrease compared to regular petroleum diesel.

Biodiesel, produced from different vegetable oils, seems very interesting for several reasons: it can replace diesel oil in boilers and internal combustion engines without major adjustments; only a small decrease in performance is reported; almost zero emission of sulfates; a small net contribution of CO_2 when the whole life-cycle is considered; and emission of pollutants comparable to that of diesel oil. For these reasons, several campaigns have been planned in many countries to introduce and promote the use of biodiesel (Carraretto et al. 2004).

Several results related to the influence of micronutrition on yield of oil have been reported on the impacts of changes in soil (Nithedpattrapong et al. 1995; Tilman et al. 2006; Thamsiriroj 2007; Thamsiriroj and Murphy 2009). Suitable climatic and soil conditions have increased plant oil yields. Biomethane generated from grass requires four times less land than biodiesel from rapeseed to produce the same gross energy. Grass is a low-energy input crop and it is a carbon-negative crop.

References

ARS. 2002. Air Resources Board, California Environmental Protection Agency. Study of CNG and diesel transit bus emissions. Air Resources Board, 2002.

Balat, M., Balat, H. 2008. A critical review of bio-diesel as a vehicular fuel. Energy Convers Manage 49:2727–2741.

Carraretto, C., Macor, A., Mirandola, A., Stoppato, A., Tonon, S. 2004. Biodiesel as alternative fuel: Experimental analysis and energetic evaluations. Energy 29:2195–2211.

Corro, G., Ayala, E. 2008. Bioethanol and diesel/bioethanol blends emissions abatement. Fuel 87:3537–3542.

Coronado, C.R., de Carvalho, J.A., Silveira, J.L. 2009. Biodiesel CO_2 emissions: a comparison with the main fuels in the Brazilian market. Fuel Proc Tech 90:204–211.

Demirbas, A. 2006a. Theoretical heating values and impacts of pure compounds and fuels. Energy Sources Part A 28:459–467.

Demirbas, A. 2006b. Global biofuel strategies. Energy Edu Sci Technol 17:27–63.

Demirbas, A. 2007. Modernization of biomass energy conversion facilities. Energy Sources Part B 2:227–235.

Demirbas, A.H. 2009. Inexpensive oil and fats feedstocks for production of biodiesel. Energy Edu Sci Technol Part A 23:1–13.

Dorado, M.P., Ballesteros, E.A., Arnal, J.M., Gomez, J., Lopez, F.J. 2003. Exhaust emissions from a diesel engine fueled with transesterified waste olive oil. Fuel 82:1311–1315.

Dunn, R.O. 2001. Alternative jet fuels from vegetable-oils. Trans ASAE 44:1151–757.

EPA (US Environmental Protection Agency). 2002 A comprehensive analysis of biodiesel impacts on exhaust emissions. Draft Technical Report, EPA420-P-02-001, October 2002.

Graham, L.A., Belisle, S.L., Baas, C.L. 2008. Emissions from light duty gasoline vehicles operating on low blend ethanol gasoline and E85. Atmospheric Environ 42:4498–4516.

Granda, C.B., Zhu, L., Holtzapple, M.T. 2007. Sustainable liquid biofuels and their environmental impact. Environ Progress 26:233–250.

Grassi, G. 1999. Modern bioenergy in the European Union. Renew Energy 16:985–990.

Janulis, P. 2004. Reduction of energy consumption in biodiesel fuel life cycle. Renew Energy 29:861–871.

Jenkins, B.M., Baxter, L.L., Miles, Jr, T.R., Miles, T.R. 1998. Combustion properties of biomass. Fuel Process Technol 54:17–46.

Kim, H., Choi, B. 2008. Effect of ethanol-diesel blend fuels on emission and particle size distribution in a common-rail direct injection engine with warm-up catalytic converter. Renew Energy 33:2222–2228.

Krahl, J., Knothe, G., Munack, A., Ruschel, Y., Schröder, O., Hallier, E., Westphal, G., Bünger, J. 2009. Comparison of exhaust emissions and their mutagenicity from the combustion of biodiesel, vegetable oil, gas-to-liquid and petrodiesel fuels. Fuel 88:1064–1069.

Lapuerta, M., Armas, O., Herreros, J.M. 2008. Emissions from a diesel-bioethanol blend in an automotive diesel engine. Fuel 87:25–31.

Nithedpattrapong, S., Srikul, S., Korawis, C., Onthong, J. 1995. Influence of N, P, K and Mg on yield of oil palm grown on Kohong soil series. Thai Agricul Res J 13:164–174.

Rakopoulos, D.C., Rakopoulos, C.D., Kakaras, E.C., Giakoumis, E. G. 2008. Effects of ethanol-diesel blends on the performance and exhaust emissions of heavy duty DI diesel engine. Energy Convers Manage 49:3155–3162.

Shay, E. G. 1993. Diesel fuel from vegetable oils: status and opportunities. Biomass Bioenergy 4:227–242.

Soltic, P., Edenhauser, D., Thurnheer, T., Schreiber, D., Sankowski, A. 2009. Experimental investigation of mineral diesel fuel, GTL fuel, RME and neat soybean and rapeseed oil combustion in a heavy duty on-road engine with exhaust gas after treatment. Fuel 88:1–8.

Thamsiriroj, T. 2007.Optimal biomass and technology for production of biofuel as a transport fuel. Master's Thesis, University College Cork, Cork, Ireland.

Thamsiriroj, T., Murphy, J.D. 2009. Is it better to import palm oil from Thailand to produce biodiesels in Ireland than to produce biodiesel from indigenous Irish rapeseed? Appl Energy 86:595–604.

Tilman, D., Hill, J., Lehman, C. 2006. Carbon-negative biofuels from low-input high diversity grassland biomass. Science 314:1598–600.

Tzirakis, E., Karavalakis, G., Zannikos, F., Stournas, S. 2007. Impact of diesel/biodiesel blends on emissions from a diesel vehicle operated in real driving conditions. SAE Technical Paper. SAE International, Warrendale, PA.

Yu, S., Tao, J. 2009. Simulation based life cycle assessment of airborne emissions of biomass-based ethanol products from different feedstocks planting areas in China. J Cleaner Prod 17:501–506.

Index

MIX
Papier aus verantwortungsvollen Quellen
Paper from responsible sources
FSC® C105338
FSC
www.fsc.org

If you have any concerns about our products,
you can contact us on
ProductSafety@springernature.com

In case Publisher is established outside the EU,
the EU authorized representative is:
Springer Nature Customer Service Center GmbH
Europaplatz 3, 69115 Heidelberg, Germany

Printed by Libri Plureos GmbH
in Hamburg, Germany